Penguin Handbooks

The Alternative Printing Handbook

Chris Treweek studied textile design, and while working at Inter-Action Trust in north London she developed the use of screen printing as a community resource. After a post-diploma course in graphic design at St Martin's School of Art, she worked part-time at the Islington Bus Company for five years, training groups to produce their own posters. She is now a freelance graphic designer. She has also written and published a set of four wall charts, *How to Screen Print*.

Jonathan Zeitlyn lives and works in north London. He has been active in campaigning and community groups, and has been involved with a number of community newspapers. He was trained and has worked as a printer and designer and as a teacher, and is now a community-education worker. His publications include *Print: How You Can Do It Yourself*, a handbook for community groups in Britain. He is also preparing a handbook for Third World development projects entitled *Low Cost Printing for Development*.

The Islington Bus Company is an educational charity set up in 1972 to explore the social and recreational needs of Islington and to establish and support self-help/non-commercial, non-party-political groups in the area. It provides training and information, loans resources such as discos, projectors, videos and slide-shows, and also makes available at no charge facilities that include duplicating, a silk-screen workshop and dark-room space. Currently the Bus Company is preparing a book compiled of all its training, activity and craft leaflets. The Bus Company is run by a mixed collective of eight, who share decision-making and general administration.

The Alternative Printing Handbook

Chris Treweek and Jonathan Zeitlyn
with the Islington Bus Company

Line illustrations by Andy Ingham

Penguin Books

Penguin Books Ltd, Harmondsworth,
Middlesex, England
Penguin Books, 625 Madison Avenue,
New York, New York 10022, U.S.A.
Penguin Books Australia Ltd, Ringwood,
Victoria, Australia
Penguin Books Canada Ltd, 2801 John Street,
Markham, Ontario, Canada L3R 1B4
Penguin Books (N.Z.) Ltd, 182–190 Wairau Road,
Auckland 10, New Zealand

First published 1983

Copyright © Chris Treweek and Jonathan Zeitlyn
with the Islington Bus Company
All rights reserved

Typeset by CCC in Univers Medium
Printed and bound in Great Britain by
William Clowes (Beccles) Limited, Beccles and London

Except in the United States of America, this book is
sold subject to the condition that it shall not, by
way of trade or otherwise, be lent, re-sold, hired out,
or otherwise circulated without the publisher's prior
consent in any form of binding or cover other than
that in which it is published and without a similar
condition including this condition being imposed on
the subsequent purchaser

Contents

Preface 7

Introduction 9
Choosing a printing method 9
Distribution 10

Stencil Duplicating 13

Screen Printing 33

Offset-litho 61

Other Printing Methods 71
Photocopying 71
Relief printing 74
Rubber stamps 78
Letterpress 79
Spirit duplicating 82

Design and Paste-up 85

Glossary 106

Useful books 109

Organizations 110

Acknowledgements 111

This book is dedicated
to our children,
Annie, Jack and Benjamin

Preface

This book has grown out of our experiences of working with many groups and individuals, from whom we have learnt a great deal. We should like to thank all of them, in particular those groups that have let us use their work as examples (⇨ Acknowledgements, page 111).

The book has taken a long time to write and we should also like to thank the many people who have contributed their ideas and criticisms during its progress. Special thanks go to Stuart Allardyce and the rest of the Islington Bus Company for all their hard work on the chapter on stencil duplicating and on the section on distribution. Our thanks go as well to Nick Cutler, Sam Harper, David Stracey and Silvie Turner, for technical advice; to Julia Vellacott and Michael Dover at Penguin Books, for their amazing ability to keep our noses to the grindstone; to Pat Kahn, for her many helpful comments; and to Patty Johnson, for her typing. Finally, we should like to acknowledge the four years of support and child care put in by our partners, Sushila Zeitlyn and Chris Ukleja, without which we would not have been able to write this book at all.

Chris Treweek and Jonathan Zeitlyn, December 1982

Chris Treweek

Jonathan Zeitlyn

ISLINGTON BUS CO.
PALMER PLACE, N7 8DH
tel 609 0226

Poster publicizing a community print-shop

Introduction

- *Community newsletter to produce?*
- *Jumble (rummage) sale or demonstration to organize?*
- *Festival to publicize?*
- *A political point to get across?*

If you need to print something – posters, pamphlets, leaflets, newsletters, or even T-shirts – to put your message over, you will find this book invaluable. It is a book not for craft-lovers but for people with something to say, something which needs to be printed, whether in five copies or five thousand copies; and it is intended not for professional designers or printers, but for everyone who needs access to their specialist knowledge and skills.

Most of the methods in the book are well within the scope of do-it-yourself; this is in any case the only way some groups can afford to print at all. If you do it yourself, you gain more control over what you are going to say and what it is going to look like. However, when you need to have something printed for you this book helps you to choose the right kind of printer and gives advice to help you keep costs down.

We have tried to make the book suitable for use in both the U.K. and the U.S.A., taking into account differences in printing practices; in general, we have used the British terminology. We have found that the international standard metric paper sizes in use in the U.K. are simple and convenient and we have used them in this book. The sizes in inches of the main ones are $33\frac{1}{8} \times 23\frac{3}{8}$ (A1), $23\frac{3}{8} \times 16\frac{1}{2}$ (A2), $16\frac{1}{2} \times 11\frac{3}{4}$ (A3), $11\frac{3}{4} \times 8\frac{1}{4}$ (A4) and $8\frac{1}{4} \times 5\frac{7}{8}$ (A5). The most commonly used U.S. paper sizes are $8\frac{1}{2} \times 11$ (regular letter size), $8\frac{1}{4} \times 14$ (legal size) and (the less common wide sheet) 11×14. Further information on sizes and weights is given in the Glossary (⇨ page 106). It should be noted that, in the U.K., the symbol # means space (in typography); in the U.S.A. it also means number. We have called 'photostats', 'bromides', 'Veloxes', 'PMTs' etc. all 'photo-prints', and 'artwork', 'mechanicals', 'camera-ready layouts' etc. the 'paste-up'. Dollar prices are approximate dollar costs (not sterling equivalents) in December 1982.

Not many books are written with the help of a bus company, but the Islington Bus Company is different. It is a resource centre for community groups in Islington in north London. It has only one bus, but lots of duplicating and screen-printing equipment.

We have written from our own experience of working with community groups in the U.K., but we hope that the information will be useful to groups working in any context or country as well as to individuals. We haven't told you everything there is to know, but enough information is provided to start you off . . . your skills will only develop with practice.

CHOOSING A PRINTING METHOD

This book covers quite a range of printing methods, but the three main chapters describe the methods that we have found the most useful, accessible and easy to use for non-commercial groups: stencil duplicating; screen printing; and offset-litho printing. All the printing methods can be carried out by you yourself, but, in practice, with offset-litho the printing is usually done for you. However, with offset-litho, as with the other photo methods, it is very easy to make the original (the paste-up) yourself, and we have described how to do this in the Design and Paste-up chapter (⇨ page 85).

Choosing the right method for the job you need to do may save your time, your money and/or your patience, so it's worth spending a little while thinking about this.

The table overleaf is a very brief guide to the main things to consider when deciding which method suits your needs best. For more detailed information on when to use each method you must read the relevant chapter.

Choosing a method

Method	Size of print	Number	Cost	Access	DIY?	How quick?	Print quality Photographs	Print quality Solids	Colour	Registration
Stencil duplicating	A4, foolscap, some A3	50–3,000	£3.50 ($7): 500 × A4 (both sides)	Good	Yes	30 minutes	Difficult	Limited results	Yes	Poor
Offset-litho	Small offset: A4 and A3	50+	£18 ($36): 500 × A4 (both sides)	Good	Not usually	2–3 days	Fair	Fair	Yes, but expensive	Good
	Larger printers: A2+	500+	£80 ($160): 1,000 × A2	Fair	No	1–2 weeks	Good	Good	Yes, but expensive	Good
Screen printing	Any size	10–200	200 posters: £15–£20 ($30–$40)	Fair	Yes	½–1 day	Possible but more expensive	Good	Yes	Yes (depends on skill)
Photo-copying	A4, foolscap, some A3	1–50	4p–10p (10¢–20¢) each	Good	Yes (easy)	Quick	Difficult	Difficult	Some full-colour, but expensive (80p–£1/$1.60–$2 each)	—
Spirit duplicating	A4, some A3	5–200	50p ($1): 100	Fair	Yes	10 minutes	Difficult	Difficult	Yes	Not necessary
Relief printing	Up to A4, depending on press	1–50	—	—	Yes	Laborious	—	Depends on press	Yes	Difficult
Rubber stamps	Three inches square	As many as you need	£5–£15 ($10–$30)	Fair	Yes	Laborious	—	No	Limited	No
Hand letterpress	Usually up to A4	10–1,000	Just the paper and ink	Limited	Yes, but a skilled job	Quite time-consuming	Expensive	Yes	Yes	Yes
Commercial letterpress	Any size	100+	Expensive	Limited	No	2–3 weeks	Yes	Yes	Yes	Yes

DISTRIBUTION

What is critical when you plan your publication is that you should carefully work out the number of copies you think you will need, as this may affect the production method you use. You must also think carefully about distribution: how you are going to get your product into the hands of the target readership? Ending up with piles of unwanted copies or falling far short of demand are equally unhappy positions to find yourself in. Experience will help you to judge better: always make a note of how many copies you used (or did not use) so that if you miscalculated you may decide more wisely next time.

PUBLICITY FOR EVENTS

Leaflets can be put through doors on a street-to-street basis. This involves hard footwork, but it can be shared if the task is divided up by allocating a street or block to each person in the group or to someone from that street. Leaflets can be displayed in local shop-windows or on notice-boards in libraries, community centres and other public places, where they can also be left in piles for people to help themselves.

Handing out leaflets in the street is quite effective if the spot is chosen with care. If appropriate, try outside your local station, factory or shopping centre, or at festivals or demonstrations. You could try to get your leaflet given out by the milkman or another travelling trader; or try to get a local shop to give them away with each purchase. An anarchist greengrocer in London did this by wrapping the vegetables in anarchist literature! However you distribute your leaflets, try to think carefully about how many you actually use – it is very easy to overestimate your needs and you will realize this when you count up the number of places in which you will be able to display them.

Posters can be displayed in similar places to the leaflets. But you will often find that if the poster is too large some places will not be able to display it. When you go round distributing it is a good idea to take along tape and pins and put up the posters yourself (after getting permission to do so). Then you know that they have in fact been displayed – institutions often say 'yes' to a poster but never actually get round to putting it up.

Fly-posting can have an impact; again, the choice of site is important both for the message to be read and to emphasize a point. One group that was campaigning about empty houses covered each house in question with a poster which named the offending owners. Fly-posting or billsticking if used with discretion can be effective and reach a wide audience.

When publicizing an event try to get some feedback: see how many people found out about it – so that you can work out what publicity works best and improve it next time.

LOCAL PUBLICATIONS

For a small-scale publication with a local audience, like a newsletter, you must make an early decision about whether to sell it or whether to give it away free. Selling helps to cover the printing costs and gives buyers an incentive to read the publication, but takes much time to organize and may reduce the number of people who will read the publication. If you give the publication away and distribute it to every local household you may reach a large audience – but you have no guarantee that everyone will read their copy! The cost of producing something like a newsletter could be met from ads, by running a jumble (rummage) sale, or simply paid for from the group's funds.

DISTRIBUTION 11

MAILINGS AND POSTAGE

Leaflets, newsletters and the like are often sent out regularly to a mailing list of people who want to be informed of a group's events. Mailings are probably very efficient as a way of getting the information to the right people and places, but the main drawbacks are the work and the cost. Addressing, filling and stamping envelopes is time-consuming, and the cost of envelopes and postage can be quite high. Joint mailings with other organizations will save time and money and may also get your publication to a wider audience.

The weight of the paper you print on may significantly affect the cost of postage – it may therefore be worth printing on both sides of the paper or using lighter paper to cut down on the total weight of the posted envelope. For example, a twenty-four page A4 booklet with cover and envelope will weigh under 100 g (3½ oz) when printed on 70 g/m² (45 lb) paper, but on 80 g/m² (55 lb) it will weigh 105 g (4 oz) and thus fall (in the U.K.) into a higher postage charge rate. So be careful when planning to print material that is to be posted.

SELLING

Street selling. Publications intended for a wide audience can be sold in the street, at work, and at other places where the public can be met like pubs or stations. It may be better if you do the selling at a regular time and place, or during demonstrations, festivals or other special events. This is very time-consuming, but it can be a good way of getting your publication to a wider readership. If your group has many committed supporters, you will be able to find volunteers to sell. Otherwise you may need to offer the sellers a discount so that they may each earn something for every copy they sell.

Shops. Newsagents, bookshops and news-stands may take your publication, but they usually want a 33 per cent discount and to take the publication on a 'sale or return' basis, which means that they only pay you for what they sell, returning the rest to you. (Note that the publication may not sell well if it is not displayed well: you can help here by ensuring that your publication is designed with a striking visual cover on which the name features clearly.)

Other shops and organizations may take your publication on the same terms. Try approaching stores, advice or community centres, or other suitable places.

Organizing a stall to sell your own and related publications is a good way of selling without having to use shops. It can also give you a direct contact with your potential readership. Try setting up a stall at any relevant event, like a festival, rally, jumble (rummage) sale or conference.

Distributors. Commercial distributors, who take books and magazines to distribute to newsagents and bookshops, deduct quite a high percentage of the cover price (50 to 60 per cent or more), but get round to more outlets on a regional or national scale than you could reach yourself. Censorship, however, may be imposed by these distributors who, for fear of libel or because they do not agree with their politics, will not take some publications.

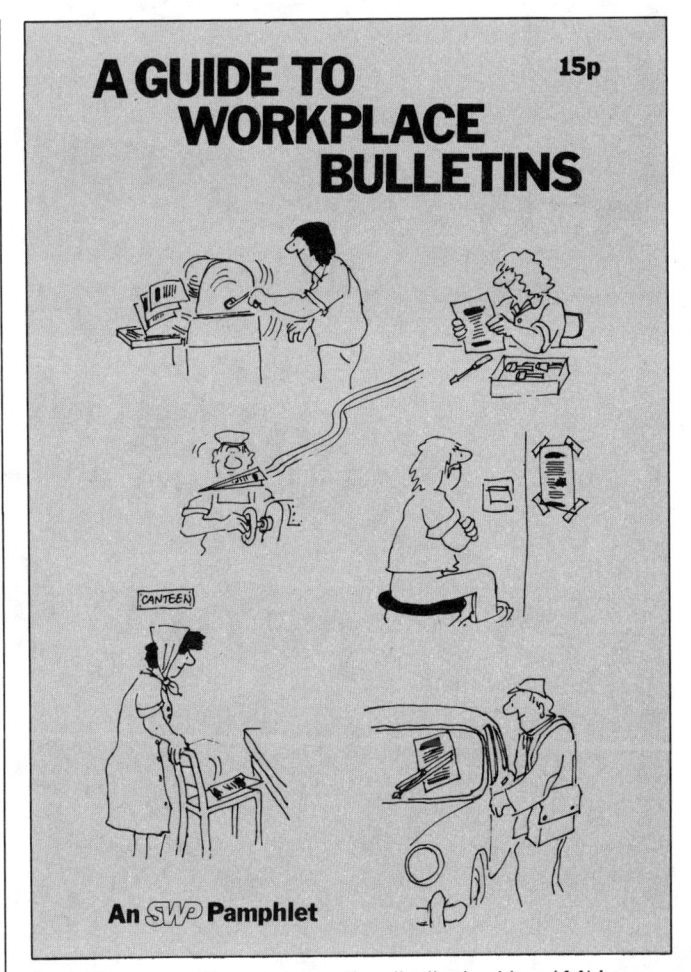

Pamphlet cover with some interesting distribution ideas (A4)!

To avoid such problems, publishers have got together to form cooperative distributors like the Publications Distribution Cooperative in the U.K. This co-op distributes left-wing and independent books, pamphlets and magazines to almost all the left, community and alternative bookshops and to some whole-food shops. It has enabled many small radical publishers to get their publications into shops that otherwise would not have stocked them. PDC, who take 52½ per cent of the cover price, have enabled the publications they serve to achieve a much wider distribution.

In the U.S.A. there are alternative distributors such as Carrier Pigeon in Boston, and Bookpeople in Berkeley. Their rates may be as high or even slightly higher than those of commercial distributors, but they place publications that no large commercial operation will accept.

Other publishers that have collaborated to help each other to distribute their books on a national scale include specialist publishers such as a local community or history or cookery-book publishers.

These examples may be worth following for other types of publications and in a more local context.

mayville puppet festival

MAYVILLE COMMUNITY CENTRE
June 27th–July 2nd
CHILDREN'S
PERFORMANCES
Mon. June 27th

punch & judy
6.30pm
Admission 10p   5–11 year olds

AND

Wed. June 29th
The Travelling Booth Co.
6.30pm
Admission 10p
5–11 year olds

AND

Derek Carpenter's
PUNCH & JUDY
ON
Sat. July 2nd
Admission 10p   5–11 year olds

AND

cap & bells
2.00 & 3.30pm

dala PUPPETS

Leaflet using electronically cut stencil (A4)

Stencil Duplicating

Introduction 13
How does it work? 13
When to use stencil duplicating 13
Finding a duplicator 13

Stencils 14
Hand-cut stencils (typed, drawn, brush) 14
Electronically cut stencils 16
Using a scanner 16
Making stencils for printing more than one colour 18
Printing smaller than A4 19
Gluing stencils together 19

The duplicating machinery 20

Operating a stencil duplicator 21
Setting up the machine 22
Putting on the stencil 22
Loading the paper and printing trial copies 23
Final printing 25
Removing the stencil 25
Things that can go wrong 25
Ink and paper 26
Special operations (printing on both sides of the paper, changing the colour, printing in two or more colours) 26
Cleaning and maintenance 27

Setting up and running a duplicating print-room 28
Arranging the room 28
Training 29
Day-to-day running 29
Managing the print-room 30

The flat-bed duplicator 31
Using a flat-bed duplicator 31

INTRODUCTION

Stencil duplicating is a quick method of printing on to A4 ($11\frac{3}{4} \times 8\frac{1}{4}$ in.) or foolscap ($13\frac{1}{2} \times 8\frac{1}{2}$ in.). The machinery is widely available, and if you are able to do the work yourself it is very cheap, because you usually only have to pay for the materials. You can also have stencil duplicating done for you, but this is not really worth while since it is so much cheaper to do it yourself, and for a little extra expense you could get better quality by using litho.

Traditionally, stencil duplicating is recognized by its blotchy, badly printed and illegible typed sheets of information – but it doesn't have to be like this! In this chapter we want to show you that it is also possible to duplicate interestingly designed, well-printed and colourful things too.

HOW DOES IT WORK?
A stencil duplicator is a machine which forces ink through a stencil on to absorbent paper. The stencil, which can be cut in a variety of ways (typed, drawn or electronically cut), is clipped around the drum of the duplicator. As each sheet of paper is fed through the machine it is pressed against the stencil, and this forces the ink through the holes in the stencil on to the paper.

WHEN TO USE STENCIL DUPLICATING
Duplicating is often used for simple typed sheets (especially in offices), but it is also very useful for printing newsletters, leaflets, small posters, tickets or cards.

Size. Most duplicators print on to A4 ($11\frac{3}{4} \times 8\frac{1}{4}$ in.) or foolscap ($13\frac{1}{2} \times 8\frac{1}{2}$ in.) paper. Machines for A3 ($16\frac{1}{2} \times 11\frac{3}{4}$ in.) and larger formats are made but are much less common. Of course, it is also possible to print smaller sizes (⇨ page 19).

Run (number of copies). From one stencil it is possible to make thousands of copies; in practice the stencil is likely to be ripped or blotched at around 5,000 unless you are experienced, careful and lucky. You can keep cutting new stencils to carry on printing to complete the job but this will take a long time and will increase the cost. If you need 2,000 or more copies get a quote from an offset-litho printer to compare costs.

Quality. Stencil duplicating is not a good method if you need very high-quality prints; large areas of solid colour or fine halftone (⇨ Glossary) do not print well. Also, on the absorbent paper the prints look a bit grey and not very crisp, and you tend to get 'show-through' (printing visible on the back of the paper), which is a problem if you want to print on both sides.

All the same, good-looking prints *are* possible if you take these limitations into account when you design for duplicating (⇨ page 16).

Time. It is much cheaper to do things yourself, but it does take time. If you are busy you may prefer to spend more money and get it done for you, or use another process.

FINDING A DUPLICATOR
Access to duplicators is very uneven; it depends on where you live. Apart from the commercial companies that do the duplicating for you (like office service shops and stationers), many big cities have a non-commercial resource centre or a community print-shop where you can use equipment, usually paying simply for the cost of the materials used.

Other organizations like those running community projects, community centres, and voluntary groups often have equipment that they let other groups use. It is also worth approaching bodies

Festival programme cover shows the difficulties of printing large solids by stencil duplicating (A5)

like schools, adult-education institutes or continuing-education centres, libraries, and local-government departments to see if they will let you use their equipment. Maybe no one has asked them before and you will pave the way for other groups! Make sure that you use the equipment carefully, clean up afterwards and report any faults in the machinery. If you have trouble finding a duplicator, ask other groups where they do their printing (look to see if this is written on their publications), or speak to a friend who works in an office!

You might also consider buying your own equipment or sharing it with another group. If you allow other groups to use your equipment (and we hope you will) make sure it is readily available and is properly looked after (⇨ page 28 for more about running a duplicating print-room).

STENCILS

A duplicating stencil is made from a sheet of plastic or coated paper. Holes are made in the stencil in various ways, and ink is pushed through them to make the print. A paper backing sheet protects the stencil until you are ready to print, and it is then removed. The stencil is fairly delicate (but not fragile) and should be handled carefully – not folded or crumpled.

All stencils have stiff paper 'headings' which are punched with a series of holes (e.g. Gestetner) or slots (e.g. Roneo or A. B. Dick) used to fasten the stencil to the duplicating drum. You must make sure that you use a stencil that will fit the duplicator you are going to use. You can also buy stencils which have a combined pattern that can be used on either system – this is called a 'universal heading'.

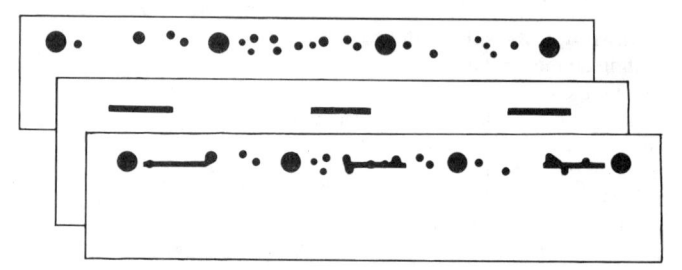

From top to bottom:
Gestetner heading, Roneo heading, Universal heading

Stencils can be bought in boxes of fifty from the suppliers, but you can usually buy the number you need from the group whose equipment you are using. There are two main types of stencil:
- Hand-cut stencils.
- Electronically cut stencils.

HAND-CUT STENCILS (TYPED, DRAWN, BRUSH)
Hand-cut stencils are used for simple typed or drawn work – the stencil is cut by typing or drawing straight on to it. Some of these stencils have a sheet of carbon between the stencil and the backing sheet so that you can see what you have drawn or typed.

Typed stencils. Before you start to type a stencil plan your page so that you know how it will look and that everything will fit in. Once the stencil has been cut it is very difficult to make anything but minor corrections. Work out on a rough any spacing you may need, both across the sheet and up and down it. For instance, if you want to type two columns of words you may want them to be evenly arranged on the page, or if you are typing headings decide how many line spaces to leave between them and the main text below, or how many line spaces between the paragraphs. All this will affect whether or not you can get all your text in – there's nothing worse than struggling down the page only to find that you don't have room for the last two sentences!

Most typing stencils have printed guidelines on them which show different paper sizes and the number of lines and characters that will fit on. When you are typing a stencil this helps you to position correctly and fit everything in. Remember that stencils are usually made to fit foolscap paper, which is 45 mm ($1\frac{3}{4}$ in.) longer than A4, so don't type right to the bottom if you are going to print on A4 paper. If you leave spaces in the typed text you can draw in any headlines or cartoons afterwards.

Drawn stencils. Drawings or hand-drawn headlines can be added on to an ordinary typing stencil, but the stencils made specially for drawing (such as Gestescript) are naturally easier to draw on and give a better-quality image.

1. *Typing the stencil*

2. *Adjusting typewriter pressure*

3. *Painting corrector fluid over typing mistakes*

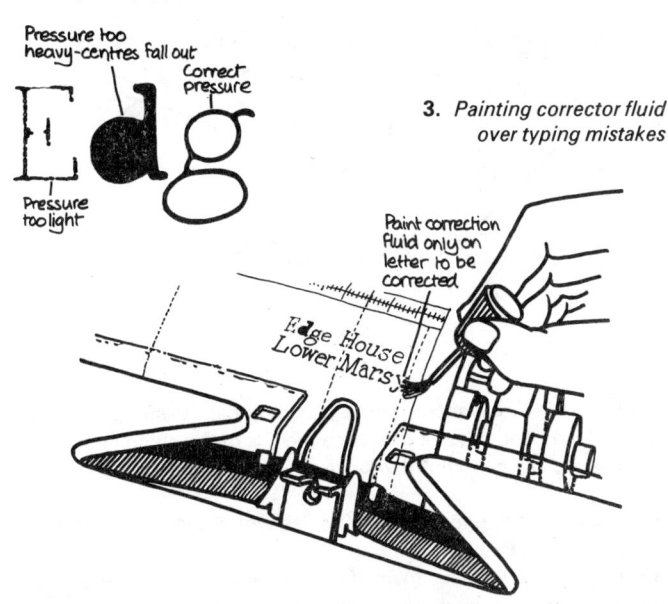

1. Put the stencil carefully into the typewriter with the stencil face up and the lines facing you. Set the colour-control knob to 'stencil' (white). This stops the ribbon coming up between the keys and the stencil, and thus makes a deeper cut. If your typewriter doesn't have a control knob of this kind simply remove the ribbon.

2. If you are using a manual typewriter, hit the keys hard and evenly so that each letter cuts well, but not so hard that the centres of the e's and o's fall out! Electric typewriters cut more evenly and some can be adjusted for extra pressure if the letters are not cutting well.

3. Before you take the stencil out of the typewriter read it through carefully to correct errors: once it is out of the typewriter it will be very difficult to make corrections. To correct any mistakes paint a little corrector fluid or nail polish on to the wrongly typed letter, allow it to dry, and then retype the correct letter.

When you have finished typing the stencil check that it has cut deeply enough by holding it up to the light – the letters should be almost clear holes. If the stencil has not cut deeply enough you will have to retype it using more pressure.

It is also possible to type on to a drawing stencil; if the bulk of your design is drawn rather than typed this may be the most suitable sort of stencil. Drawing stencils are more expensive than ordinary typing stencils. If you need to trace your design from a rough drawing (or other source) you can use a transparent drawing stencil.

First make a rough the same size as you want to print, so that you know exactly where you are going to put everything before you start drawing, remembering how difficult it is to make major corrections once the stencil is drawn. Lay the stencil down on a hard smooth surface and draw your design on it with a stylus or a ball-point pen.

Left: Leaflet for a community education project using hand-drawn stencil (A4)

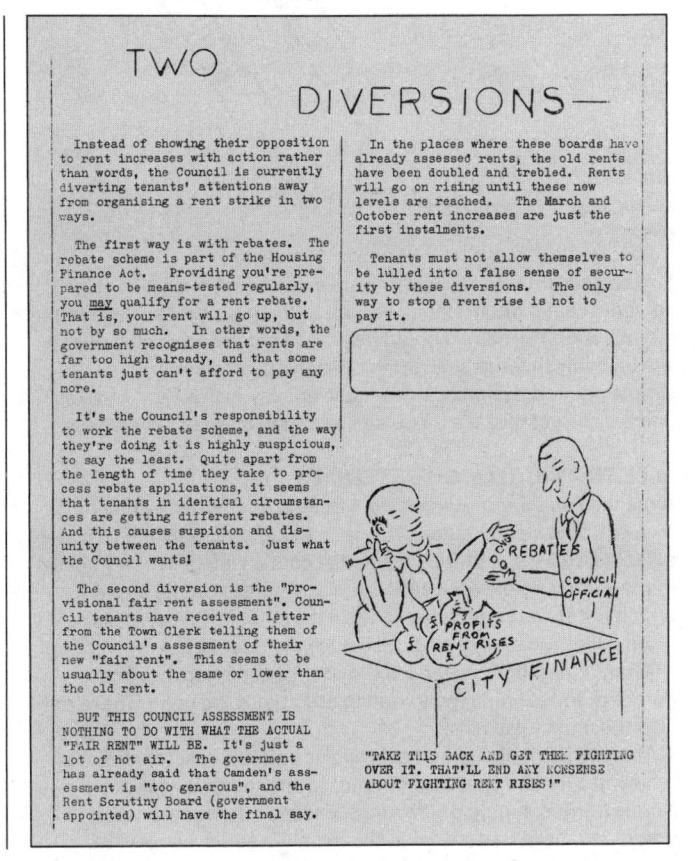

Right: Page from a tenants' newsletter using a typed and drawn stencil (25 × 19 cm)

Drawing the design on to a drawing stencil

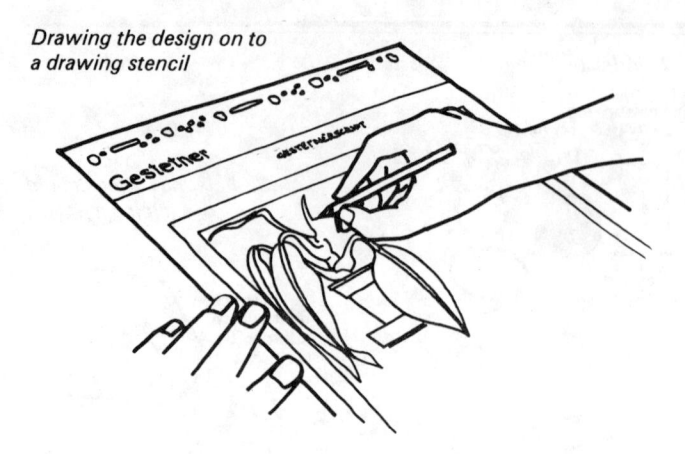

A special 'writing sheet' can be placed under the stencil to help you make a good cut and to reduce the risk of the stencil tearing. (You especially need this if you are drawing on to a typing stencil.)

There are many gadgets made for use with hand-cut stencils. The most useful are:
- *A wheel pen.* To make rows of dotted lines without weakening the stencil in the way that solid lines would. Areas of tone can also be made with a wheel pen.
- *A loop pen.* For italic writing.
- *A burnishing sheet* can also be used to produce areas of tone. Place it underneath the stencil and rub the stencil with a 'burnisher' (such as the bowl of a spoon or even a ball-point pen) to get a pattern of dots. A sheet of sandpaper produces a similar effect.

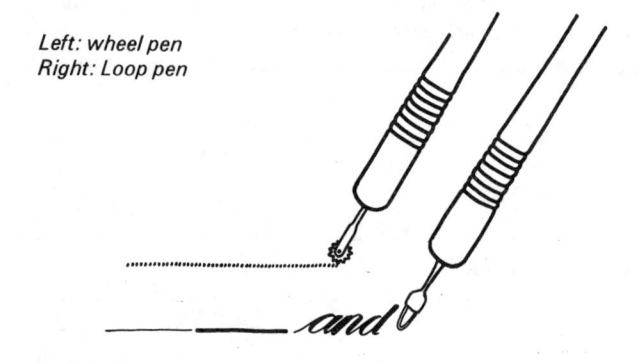

Left: wheel pen
Right: Loop pen

Brush stencils. Brush stencils are made by painting a special solvent (called brush stencil ink) on to the ink-proof surface of the stencil. Blotting-paper should be placed between the stencil and the backing sheet to soak up the dissolved-off stencil. Large simple painted designs can be made quickly in this way, but you cannot correct mistakes. You can also type on to brush stencils.

ELECTRONICALLY CUT STENCILS

A stencil can be reproduced from a piece of artwork by means of an electronic stencil cutter. This is an expensive machine (it costs at least £800 – or $1,500 – new) that cuts an image electronically in the stencil wherever it sees the black images on the artwork, which it thus reproduces exactly on the stencil. This process is often called 'scanning' and the machine a 'scanner'.

A range of different stencils is made for scanning. For finer details or longer runs the thinner plastic types are used. These are also the most expensive.

This stencil-making method enables you to duplicate designs which use instant lettering, cuttings from other publications, simple photos, typing or drawings. Paste all the components down and make a paste-up in the normal way (⇨ page 103).

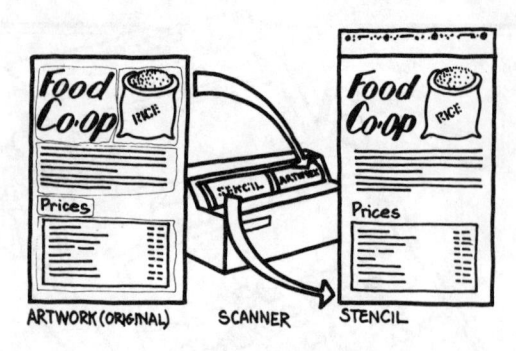

The electronic stencil cutter provides a direct method of producing a stencil from a paste-up using a variety of different elements

Special points to bear in mind when making artwork for scanning and duplicating

1. The design should fit on the A4/foolscap sheet with a border of at least 1·5 cm (½ in.) all round, because you cannot print well right up to the edge of the sheet of paper.

2. Large areas of solid image in your design will cause printing problems, because the amount of ink needed to make a good solid print will make the paper stick to the drum. An alternative is to outline areas of colour and fill them in with tone (either hand-drawn or instant tone ⇨ page 100).

3. If you need a bold heading remember to scan the artwork upside down on to the stencil, so that the solid area comes off the duplicator drum last; this will cause fewer printing problems than if the solid area were to come off the drum first (⇨ page 26). Make sure you don't have headlines at the top and the bottom or this will not be possible.

4. Because you cannot enlarge or reduce on the scanner you must make the artwork the correct size, that is, the size at which it is going to be duplicated.

5. Only photos with strong contrasts of black and white, or those which have already been screened into dots (for example, pictures from newspapers or magazines) will reproduce well.

6. Do not paste up the artwork on card that is too thick to be bent around the drum of the scanner.

USING A SCANNER
Different makes of scanners work in slightly different ways. This section describes the basic procedure but you should also consult the instruction manual for the scanner you are using; even better, get someone who knows the machine to show you how it works.

Front page of newsletter using electronically cut stencil (A4)

Annual report cover using electronically cut stencil shows how a photograph with strong contrast can reproduce well (A4)

Page from a newspaper produced by children at a playcentre using an electronically cut stencil (A4)

Leaflet for a fete using electronically cut stencil (A4)

Leaflet for a Christmas party using electronically cut stencil (A4)

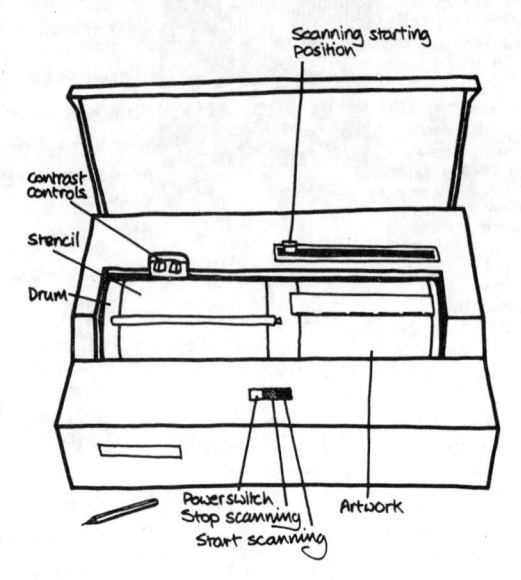

Controls on an electronic stencil cutter

The scanner is very sensitive, and if you are not careful it will pick up more than you want it to, for example, dirty marks, paste-up edges, the tone of the background paper (if you have used newspaper cuttings, say) and so on. The contrast can be adjusted on most machines so that these unwanted details are not picked up on the stencil (or, conversely, so that areas of the artwork that are not black enough are picked up more strongly). Some scanners also have a control that enables you to adjust the number of lines scanned per inch. By scanning more lines per inch you increase the quality and get more detail, but the scanning takes longer. The best approach is to test to see what settings you need by scanning an inch or so of your artwork on a test stencil. The same stencil can be used for a number of tests and this is really worth doing if it prevents you from wasting stencils . . . electronic stencils are quite expensive!

MAKING STENCILS FOR PRINTING MORE THAN ONE COLOUR

Because it is possible to change the colour of the ink in a duplicator you can produce very effective multicolour prints. You must make a separate stencil for each colour in the design, and then print the first stencil in the appropriate colour; after changing the colour of ink you print the second stencil on the same sheet of paper and so on (⇨ page 26).

Registration. Because a duplicator does not make each print in exactly the same place on each sheet, it is impossible to get very accurate registration of colours. Your design will need to take this into account – either leave a gap of about 0·5 cm ($\frac{1}{5}$ in.) between the colours, or make the colours overprint on to each other (which also produces a third colour!). Very effective multicolour designs are possible provided you plan for these and do not expect the machine to do what it is not designed to do.

Hand-cut stencils. If you are using hand-cut stencils, make a stencil for each colour in your design; the easiest way is to use transparent stencils and trace each one from the rough, or from a print of the first colour.

Scanning a stencil

1. Switch the scanner on to warm up. (Warming-up time varies with different models but is at least three minutes.) Position the artwork around the right-hand side of the drum, making sure that it is flat and that the edge is parallel to the side guideline. Clip down the plastic sheet which keeps the artwork in position.

2. Clamp the stencil in position around the left-hand side of the drum making sure that it is flat and lined up with the position line on the drum. Set the controls (if your machine has them), having made a test strip beforehand to find out the correct settings (⇨ page 23).

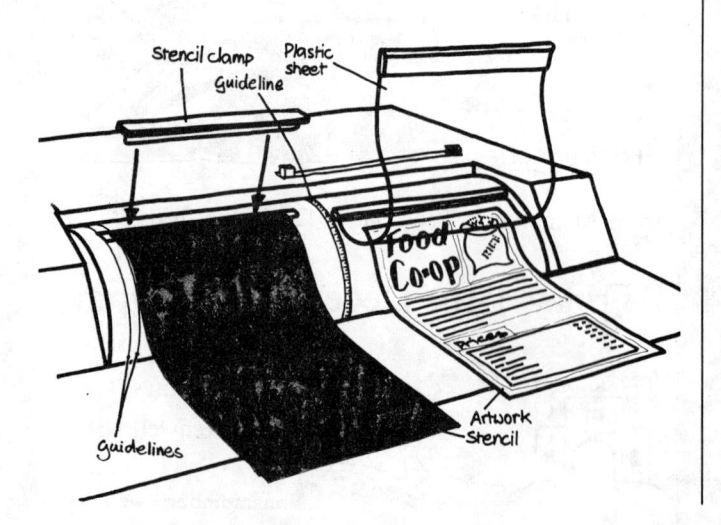

3. Make sure the photocell is in the starting position (usually across to the left) and switch on. As the drum turns, the photocell travels across the artwork and the stencil is cut simultaneously; this takes between three and ten minutes depending on the type of machine and the setting used. When the drum stops turning, remove the stencil and the artwork.

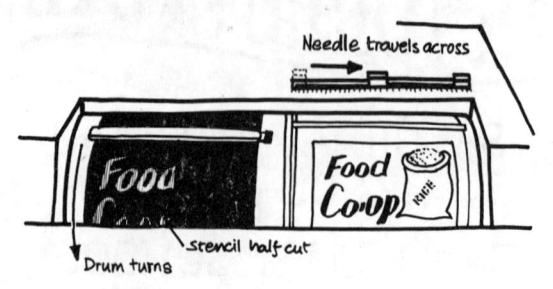

4. Over a light-box or at a window check the stencil for any unwanted marks that have picked up on the stencil, and paint them out with corrector fluid.

Electronically cut stencils. Even though you still need a stencil for each colour when you are using electronic stencils, this does not necessarily mean that you need to make a separate piece of artwork for each stencil. For example, for a simple leaflet which has only a heading in a second colour, you can make one piece of artwork and scan it twice; the first time round, cover over the heading with white paper while you make the stencil for the text, and the second time, cover over the text with paper and scan the stencil for the coloured heading.

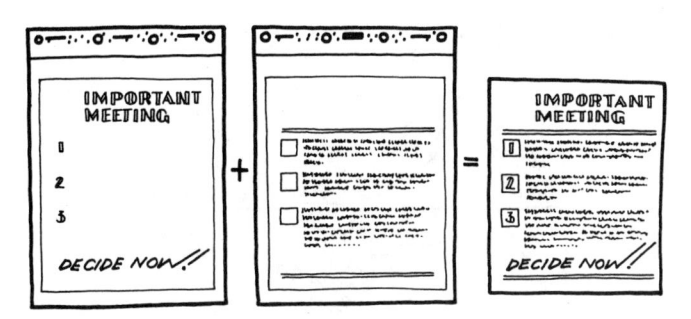

A separate stencil for each colour in the design

With more complex two-colour designs this method is not possible, and you must make a piece of artwork for each colour in the design (⇨ page 27), as follows:

1. On your rough sort out which parts of the design you want in which colour.

2. Draw out on white paper the artwork for whichever colour covers the largest area.

3. Lay a piece of tracing paper over the first artwork and draw in the design for the parts that are to be printed in the second colour. In this way you can see exactly how the two colours will link together.

4. Using the different pieces of artwork make the necessary number of stencils.

PRINTING SMALLER THAN A4
It is not easy to use a duplicator to print on to paper or card smaller than postcard size. To do so it is much better to print multiples of your small design on A4 sheets, cutting them up after they are printed. This is easiest if the divisions are simple – halves, quarters, sixths etc. The cutting can be done with a large pair of scissors or a guillotine (a paper cutter).

Hand-cut stencils. On hand-cut stencils you just draw or type out your design as many times as you need it.

Stencil with the design drawn or typed on to it four times

Electronically cut stencils. If you are scanning a stencil you can make one small original and then scan it on to the stencil as many times as you need to (this is much easier than drawing it several times), as follows:

1. Mark out four rectangles on your A4 sheet in light-blue pencil. (Light blue will not be picked up by the scanner.) Stick the artwork temporarily in position with a small blob of glue in one of the squares and scan the stencil as usual.

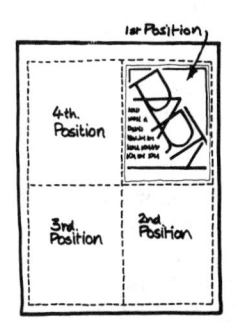

2. Leave the stencil clipped in position, and move the artwork to one of the other blue squares on the sheet. Scan it again, making sure you put the A4 sheet back in exactly the same position on the drum of the scanner. Repeat this process until you have scanned all four designs on to the stencil, then print as usual.

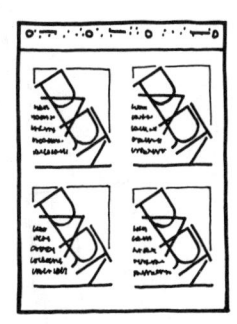

An alternative method is to use good-quality photocopies, from the original, pasted up to make an A4 piece of artwork to scan.

GLUING STENCILS TOGETHER
You can combine electronically cut stencils with hand-cut stencils by cutting them up and gluing them together with stencil adhesive. This may be useful if you use a logo or letterhead regularly; scan a number of them on to one stencil, and then use them one at a time, glued on to a typed stencil as follows:

1. Cut out the image from the electronically cut stencil.

2. Cut a hole in the typing stencil slightly smaller than the patch, so that there will be a slight overlap.

3. Glue the 'patch' in position on the top of the stencil.

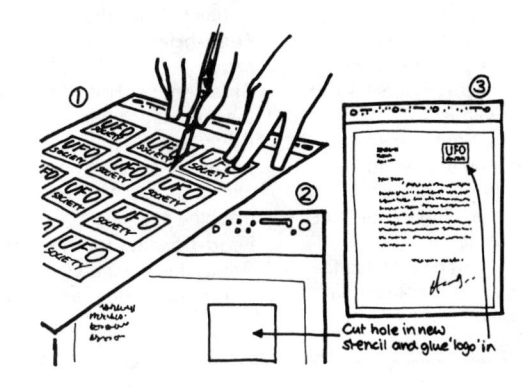

STENCIL DUPLICATING

THE DUPLICATING MACHINERY

There is a wide range of duplicators available costing from £20 ($50) second-hand to £1,000 ($2,000) for a new machine. (You can also build your own flat-bed duplicator – ⇨ page 31.) Generally, the print quality is about the same for simple hand-operated machines as for the expensive electric and automatic ones, but the latter are faster and easier to use. The advantage of the simple models is that they have fewer parts to go wrong and are easier to learn to operate (or to teach others how to use).

Most machines are based on one of two main systems: the two-cylinder type, like the Gestetner; or the one-drum type like the Roneo or A. B. Dick.

Two-cylinder-type duplicators. A duplicator of the two-cylinder type has two solid cylinders. Ink is pumped up from a tube and spread on to the cylinders as they turn by an ink distributor and inking rollers. The stencil is attached to a fabric screen which is stretched over both cylinders, and goes round as the cylinders turn. The paper is fed through the machine and pressed against the stencil by a pressure roller, and this forces the ink through the holes in the stencil on to the paper.

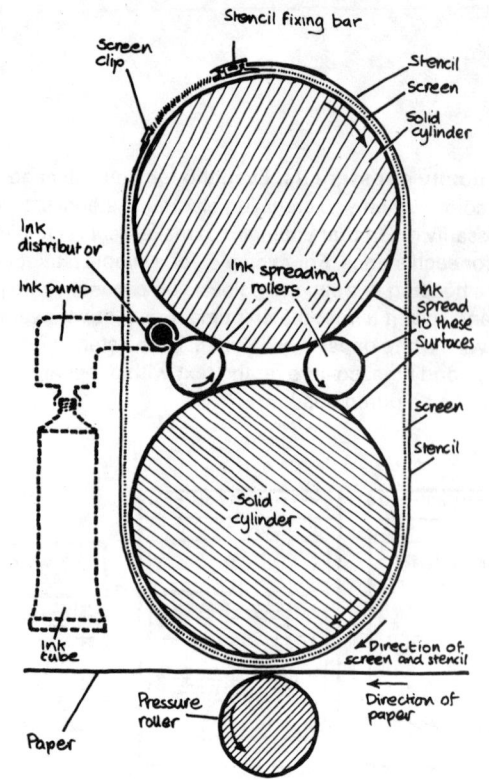

Cross-section of the two-cylinder-type system

Two-cylinder-type duplicator (Gestetner)

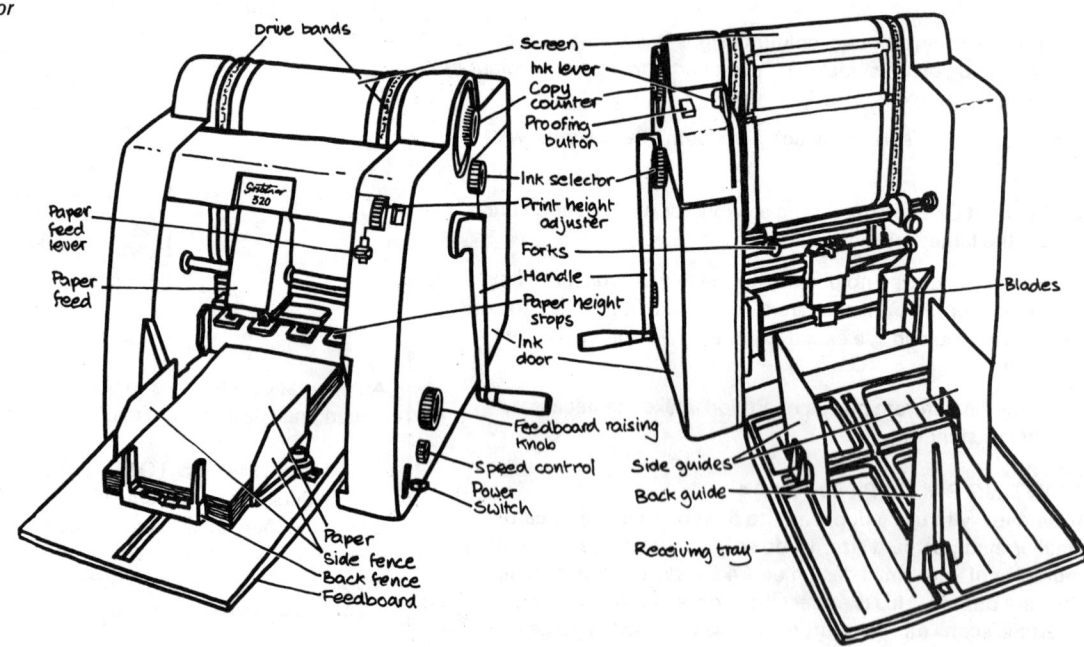

Ink tube. The ink source.
Ink pump and lever. Pumps the ink from the tube to the ink distributor.
Screen. Fine fabric stretched round both cylinders which allows ink to pass through.
Ink distributor. Ejects ink out on to the cylinders.
Ink rollers. Spread the ink evenly around the cylinders.
Stencil-fixing knobs. Fix into the stencil heading to hold the stencil in place.
Stencil-fixing bar. Holds the stencil heading in contact with fixing knobs.

Handle. For turning the cylinders and stencil by hand.
Proofing button. Brings the pressure roller up against the stencil to make the first print.
Feed-board. Holds the pile of paper.
Side fences. Hold the paper in the correct position.
Back fence. Holds a weight on top of the paper to ensure that only one sheet feeds into the machine at a time.
Feed-board raising knob. Brings the paper up to correct level to feed into the machine.

Paper-height stops. Ensure paper is raised to the correct height.
Receiving tray, side guides, back guide. Catch the printed sheets in a stack.
Print-height adjuster. Varies the vertical position of the image on the paper.
Paper feed. Feeds the paper into the machine one sheet at a time.
Paper-feed lever. Turns the paper feed on and off.
Blades, forks. Two different systems used to help separate the printed sheets from the stencil as they emerge from the machine.
Copy counter. Counts the number of copies printed.
Ink selector. Varies the supply of ink to different areas of the stencil.
Power switch. Turns on the motor.
Speed control. Adjusts the printing speed.
Pressure roller. Presses the paper against the stencil to make the print.
Drive bands. Turn the cylinders round (with the screen).

One-drum-type duplicators. A machine of the one-drum type has one large hollow drum, which is filled with ink. As the drum turns, ink comes out through small holes in the drum and soaks into the felt covering. The stencil is clipped around the drum and the felt. The drum turns and as the paper is fed through the machine it is pressed against the stencil by the pressure roller, and the ink is forced through the holes in the stencil on to the paper.

OPERATING A STENCIL DUPLICATOR

Manufacturers make it their business constantly to improve and refine their equipment and it would be impossible in this book to explain exactly how each machine works. What follows is a generalized account of how to operate a stencil duplicator, which you should be able to apply to your machine. Obviously you will also need the manual for your particular model or, better still ask someone who knows how to use it to show you. You can get manuals for second-hand machines by writing to the manufacturers, who may also send a representative who will usually be very helpful if you are having difficulty using your machine.

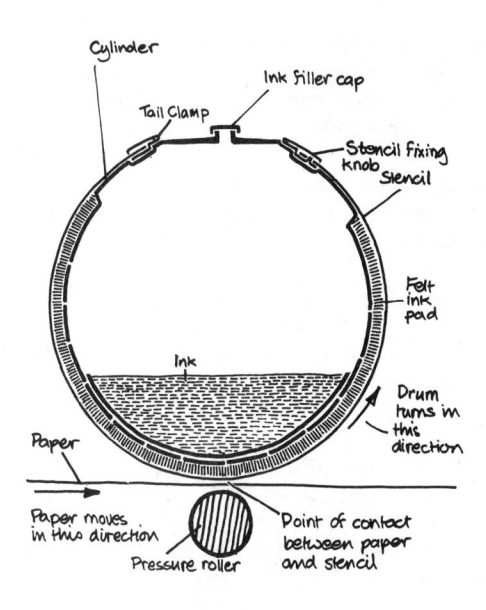

Cross-section of the one-drum-type system

One-drum-type duplicator (Roneo)

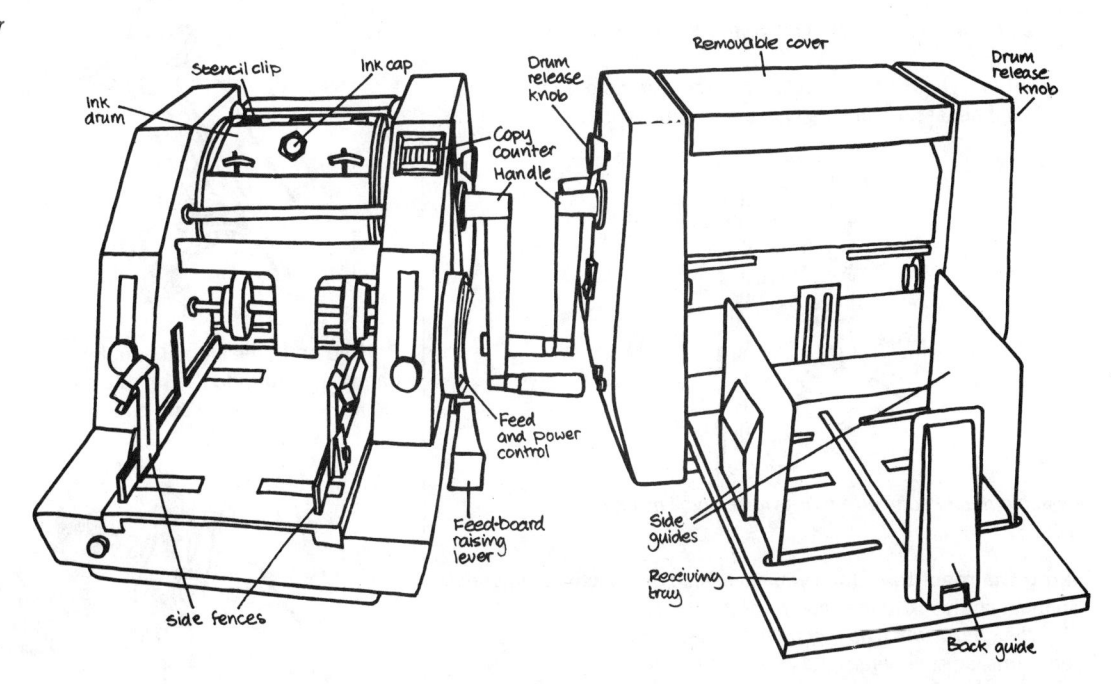

Ink drum. The large drum which holds the ink.
Ink cap. For pouring ink into the drum.
Felt pad. Covers the drum and absorbs the ink as it seeps out of the drum.
Stencil-fixing knobs. Fix into the stencil heading to hold the stencil in place.
Handle. For turning the drum by hand.
Tail clamp. Fastens the bottom of the stencil to the ink pad.
Feed-board. Holds the pile of paper.

Side fences. Hold the paper in the correct position – the weights ensure that only one sheet at a time feeds into the machine.
Feed-board raising lever. Brings the paper up to correct level to feed into the machine.
Receiving tray, side guides, back guide. Catch the printed sheets in a stack.
Print-height adjustment. Varies the vertical position of the image on the paper.
Paper-feed rollers. Feed the paper one sheet at a time into the machine.

Strippers. Separate the printed sheets from the stencil.
Copy counter. Counts the number of copies printed.
Speed control. Adjusts the printing speed.
Drum release knob. Releases drum when it is to be removed for a colour change.
Pressure roller. Presses the paper against the stencil to make the print.
Feed and power control. Turns on the motor and controls the paper feed.

STENCIL DUPLICATING

SETTING UP THE MACHINE
Here are the basic steps for using the machine, based on a Gestetner 420. (Where one-drum-type machines differ greatly the difference is described.)

Switch on. If you are using an electric machine, make sure that all the controls on the machine are in the OFF position, then connect the power supply.

Check the ink. Check that the machine has enough ink for the job:
● *On a one-drum-type machine*, pour ink into the drum through the ink cap at the top of the drum.

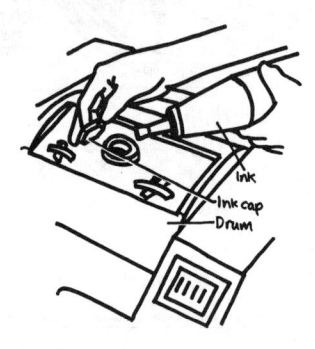

Pouring ink into the drum of a one-drum-type machine

● *On a two-cylinder-type machine*, check that there is ink in the tube and if necessary replace with a new tube of ink.

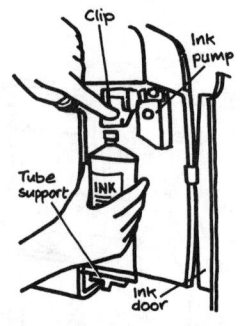

Replacing the ink tube on a two-cylinder-type machine

Inking the machine. Once you have filled the machine with ink you must apply ink to the screen (or felt pad):
● *On the one-drum-type machine*, the felt pad is inked as the drum turns, so simply turn the drum a few times.
● *On a two-cylinder-type machine*, press down the inking lever and turn the handle a few times, until the screen is shiny and evenly distributed with ink. Take care not to over-ink!

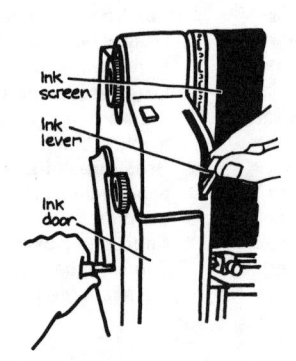

Pressing down the ink lever on a two-cylinder-type machine

Putting on the stencil

1. If you are using a hand-cut stencil, remove the carbon sheet, leaving only the stencil and backing sheet. If you are using a plastic electronic stencil, peel the backing sheet away from the stencil up to the perforations, but do not tear it off yet.

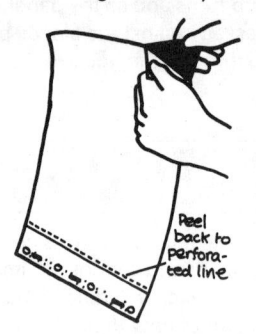

2. Place the stencil on the machine with the backing sheet facing you and the stencil on the inky screen. The screen must be shiny with ink so that the stencil will stick and stay in place. Hold the stencil with one hand and curve it slightly so that it remains stiff. Raise the fixing bar with your free hand. Place the heading of the stencil in position on the appropriate knobs or grooves, and then let down the fixing bar.

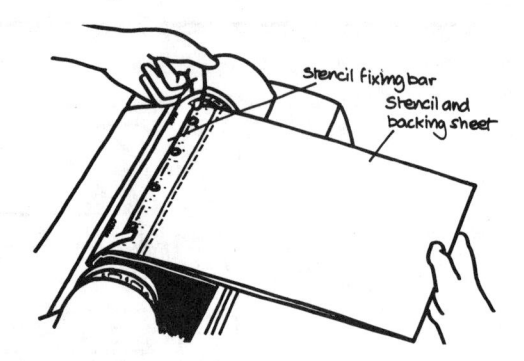

3. Hold on to the stencil and the backing sheet, and with your free hand begin turning the handle which rotates the drum. As the stencil rolls on to the screen it will stick in place.

4. When the stencil is in place smooth it gently on to the screen. Raise the backing sheet to make sure there are no creases in the stencil; if there are any smooth them out with both hands – do this carefully to avoid tearing or stretching the stencil.

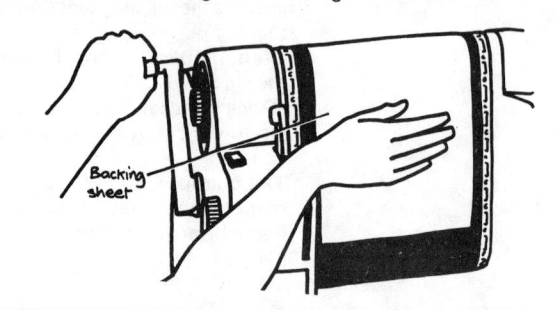

OPERATING A STENCIL DUPLICATOR   23

5. *On one-drum-type duplicators* you must now fix the bottom end of the stencil into the tail clamp. (*On a two-cylinder-type machine* this is left loose.)

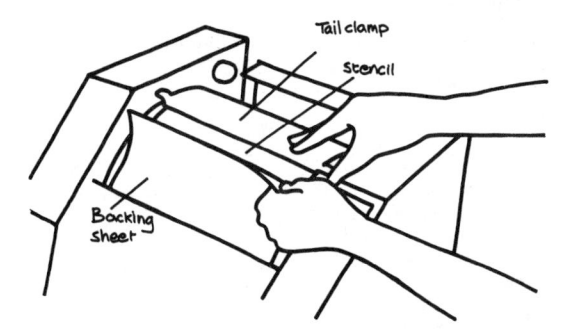

6. To test that the stencil is attached and the screen properly inked, turn the handle three or four times, press the proofing button, and then examine the backing sheet. If you are getting an evenly inked image on it and there are still no creases in the stencil you can now remove the backing sheet: if the backing sheet is not evenly inked, turn the handle a few more times and ink the screen a little more. The backing sheet has perforations, but do not rely on these alone to tear off the sheet – crease it along the perforated line and start the tear with your thumb-nail. Keep the backing sheet, you may need it later on if you are going to store the stencil.

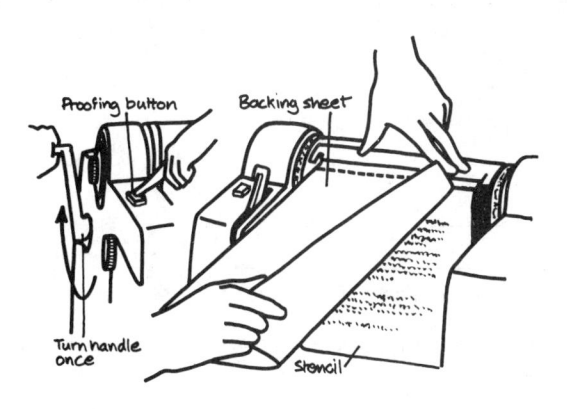

Loading the paper and printing trial copies

1. Fan the paper in case it has become damp and stuck together and then stack it in a neat pile.

2. Put the paper on the feed-board with a few sheets of scrap on the top, and fix the side and/or back fences; these should hold the paper firmly in position without being so tight that they grip it and stop it moving. The back fence has a weight which must be dropped over the paper stack. This stops the machine from taking more than one sheet at a time.

3. Raise the feed-board using the knob provided (usually located on the side of the machine) until the paper touches the paper-height stops. This puts the paper stack in the right position so that the machine can feed paper through automatically one sheet at a time.

4. On the other side of the machine is the receiving tray which catches the printed sheets of paper as they come out. It has adjustable guides to catch the paper; adjust these to catch the paper in a neat pile.

5. Set the print-height adjustment to zero.

6. Switch on the paper-feed lever, turn the handle by hand once and as you do so turn off the paper-feed lever, so that you print one copy on to the scrap paper. (Some machines have a single-copy button.)

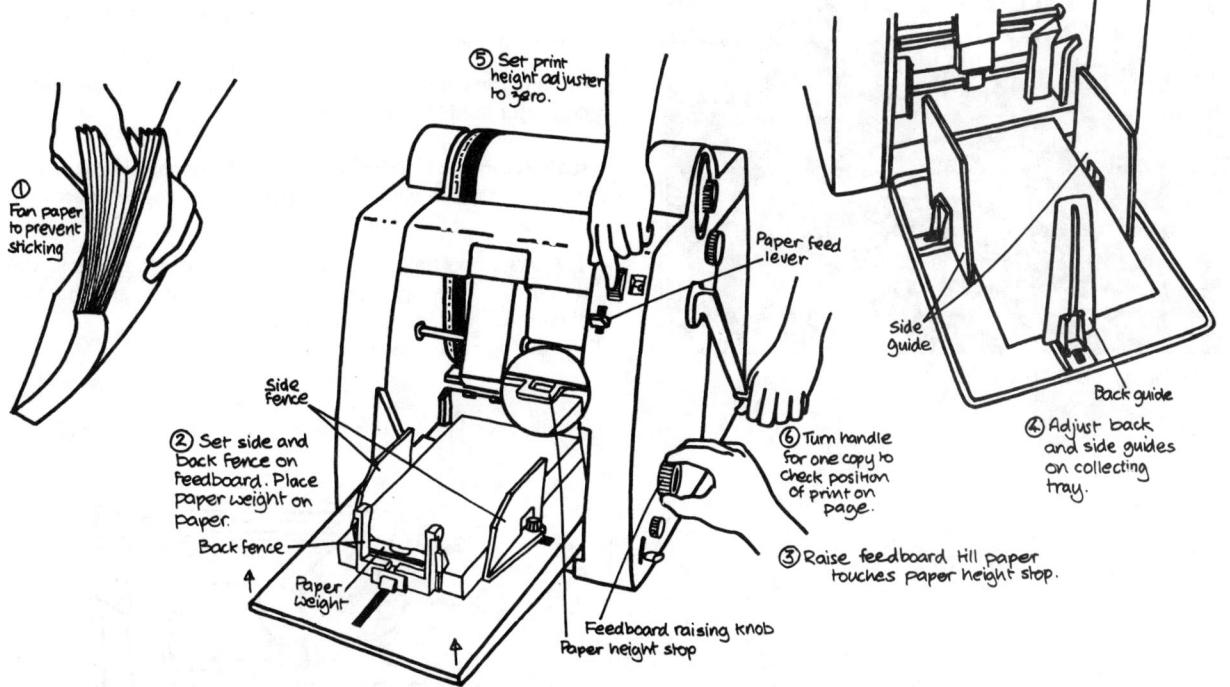

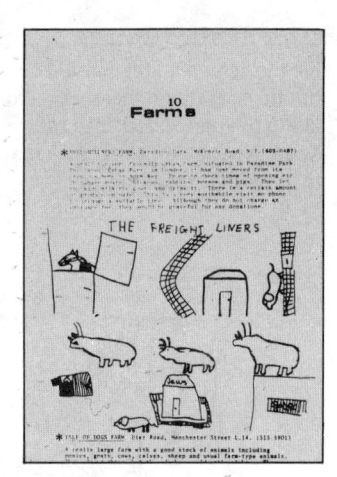

Left: Front page of a newsletter before vertical registration has been corrected (A4)
Right: Page from a pamphlet before horizontal registration has been corrected (A4)

Uneven inking on this A4 leaflet

Examine the copy you have printed for:

a. *Vertical registration.* (Is the print too high or low on the paper?)

b. *Horizontal registration.* (Is the print too far to the left or right? Are the margins even?)

c. *Evenness of inking.* (Good overall print and inking?)

d. *Unwanted marks.*

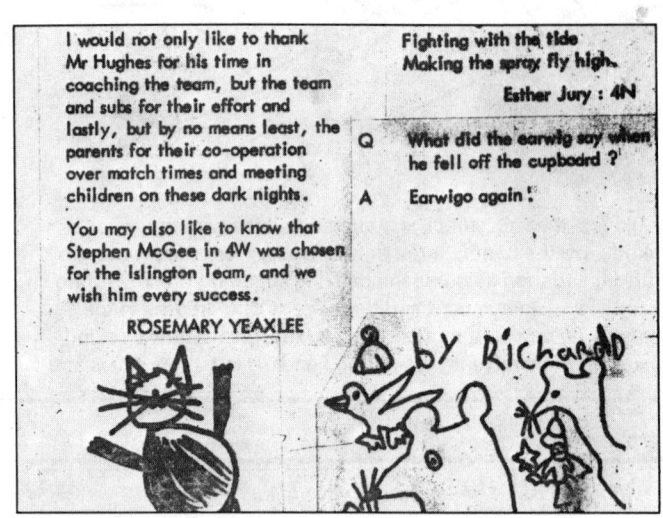

Paste-up edges have picked up on the electronic stencil cutter (detail from A5 school newsletter)

To remedy these problems you will need to:

a. Turn the print-height adjustment up or down.

b. Drop the feed-board, loosen the side fences, and move the paper pile left or right as necessary. Afterwards, put the fences back in place in their new position and raise the feed-board again.

c. Run off a few copies on to scrap paper, either to use up excess ink, or to force more ink through on to your paper. Restrain yourself from adding more ink until you are absolutely sure that this is the problem. Run off ten or twenty sheets of scrap paper and if the inking is definitely inadequate add a little more ink until you get a good print.

d. Paint over any unwanted marks or pin-holes in the stencil with corrector fluid to seal them off.

Adjusting the controls. Your machine may also have one or more of the following adjustments. Check your manual for the positioning of the controls:

1. *Forks.* Forks help separate the paper from the stencil. Make sure that they are just within the width of the paper, and that they dip to clear the cover and stencil-fixing bar as it rotates.
Never adjust forks when the machine is running.

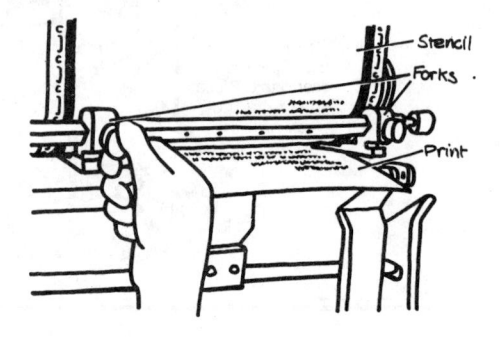

2. *Ink selector.* This governs the area of the screen to be inked. If most of your image is on one side of the stencil you do not need to ink the whole screen. A selector allows you to distribute ink to the appropriate area.

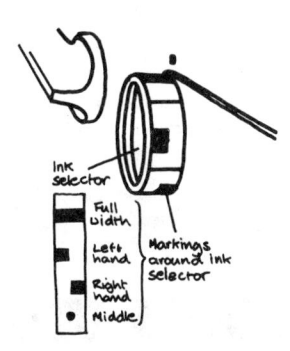

3. *Copy counter.* This counts the number of copies you print. If you do not have a counter you can make a good guess at how many sheets you have printed by remembering that there are 500 sheets in a ream.

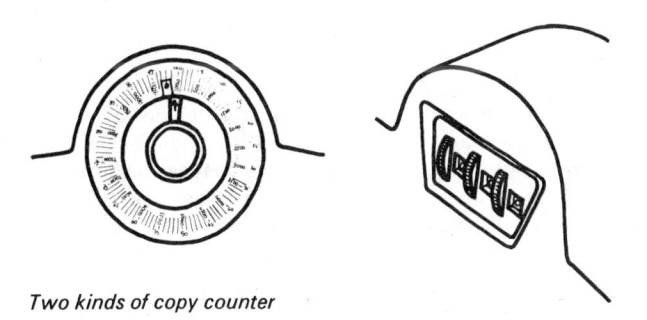

Two kinds of copy counter

FINAL PRINTING

1. If your machine is electric turn on the power button and switch on the paper-feed button. The machine will now start printing. If anything goes wrong during printing turn either of these buttons off straight away. (If you turn the power button off, but leave the paper feed on, remember to switch it off before restarting or the paper will start to feed before you are ready for it.)

2. Most machines have a speed-control knob. A slower speed allows time for the ink to soak into the paper and makes a better print.

3. Watch the copies coming off. If they are too light, first slow down the machine, and if the print is still too light you can add more ink. (Pumping up more ink won't improve the quality of a badly cut stencil – it will just cause over-inking!)
● *On a one-drum-type machine*, stop the machine and add more ink to the drum (⇨ page 22).
● *On a two-cylinder-type machine*, press down the inking lever once or twice to pump up some more ink. If this does not improve the quality stop the machine, check that there is still some ink in the tube and replace with a fresh one if necessary.

4. When your paper counter reaches zero, it should automatically stop the machine. If your machine does not do this automatically, you will have to watch the counter and turn it off manually when you have printed the correct number.

Removing the stencil

1. (On a one-drum-type machine unclip the bottom of the stencil.) Lift up the fixing bar, unclip the top of the stencil and fold back the stencil heading.

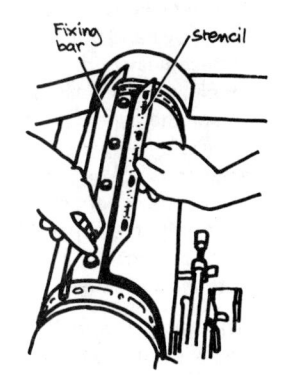

2. Remove the stencil. If you need to store it, pick up the stencil backing sheet and hold it against the inky part of the stencil.

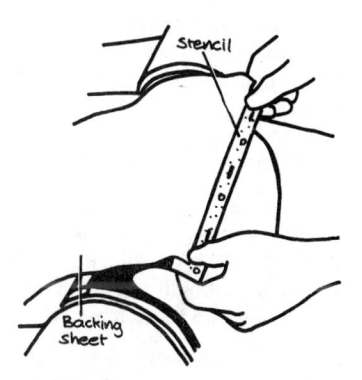

3. Holding the stencil and backing sheet pull them off the machine as you turn the handle in the opposite direction. By doing this you have put the flimsy stencil back on to its backing sheet. Now check for creases in the stencil and remove as many as possible (if left in they become permanent after a short time). The stencil can then be stored for reuse when needed. Stencils are best stored hanging vertically. Manufacturers make special cabinets and hangers for this but you can use clothes pegs or pins and a line. Do not store stencils in the sun, where they will warp and go brittle.

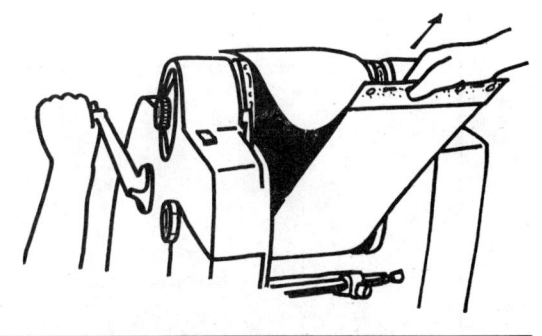

THINGS THAT CAN GO WRONG

If anything goes wrong, turn off the machine immediately and then check through the following possibilities:

1. *The paper is not feeding into the machine.* Check that:
● the paper feed is on.
● the paper weight is on.
● the paper has not been raised above the paper-height stops.
● the feed unit is correctly positioned, and the paper-feed bar moving correctly. If the rubber pads are worn they can be removed and replaced quite easily.
● the copy counter is not set on zero.

2. *The paper is not coming out of the machine correctly.* Check:
● Are the forks or blades in the correct position?

- Is the machine over-inked? This tends to make the paper stick to the stencil or get caught up in the forks. If so, run scrap paper through by hand until the surplus ink is used up.
- Is there heavy printing close to the top of the paper (for example, a big headline)? If so, the forks may not be able to get between the paper and the stencil because the large area of ink will make the paper stick to the stencil. Try lowering the image on the paper using the print-height adjustment button.
- Are the forks bent so that they are not coming into correct contact with the paper? If so, remove the bar and gently bend them back into shape – do not bend them too far or they may tear the stencil and screen.

3. *The stencil is torn.* Why?
- There is not enough ink on the screen, so pump a little more through.
- There is too much ink on the screen, so run scrap paper through by hand until the surplus ink is absorbed.
- The forks are bent too close to the screen. If so, remove and readjust.

Minor tears can be repaired with masking tape, but major tears will require a new stencil.

4. *There are ink smudges and blotches.*
- This is caused by over-inking (after all our reminders!), manifested by ink oozing out of the machine and/or getting on to the copies. Stop and clean up the mess using paper towels or rags. Make sure you remove all the excess ink.

5. *There is a shadow image on the back of the paper.*
- This is usually caused by the paper not feeding into the machine so that the pressure roller comes into contact with the stencil and then transfers the image on to the back of the paper. Run through some scrap paper until the image disappears.
- It may be caused by heavy areas of ink (due to either over-inking, or large solids in the design), which do not dry before the next sheet comes down on to the collecting tray. Try using less ink or running the machine faster or, as a last resort, interface each sheet with scrap so that each print is 'blotted'. You may be using the wrong kind of paper; if the paper is too shiny it will not absorb the ink quickly enough, in which case either change the paper or interface with scrap.

Shadow image on the back of an A5 Christmas card (detail)

6. *No ink is coming out.*
- On two-cylinder-type machines replace the ink tube.
- On one-drum-type machines fill up the drum with more ink.

7. *There is faint image on the sides of the paper.* This is because:
- The stencil is incorrectly cut. Either paint out the image with corrector fluid or, if it is really bad, recut the stencil.
- The design is too close to the sides of the paper. (You should have left a margin of at least 1·5 cm/½ in.).

- The pressure roller may be worn. It is easy to replace it with a new one: first remove the forks and control guide, then the blades, and finally disengage the roller by lifting the levers on both sides.

8. *Certain parts are not printing.*
- Check for a crease in the stencil. Ease it out and try a little extra ink on that section, then run scrap paper through until it is printing correctly.
- Check for particles between the screen and the stencil. Peel back the stencil and carefully remove any obstructions.

Creased stencil (detail from A4 playground publicity leaflet)

INK AND PAPER

Ink. This is available in many different colours – you do not always have to use black. The main thing to remember is that there are two types of ink, one for two-cylinder-type machines and one for one-drum-type machines. These inks are very different, so do not use the wrong ink for your machine – you could ruin it!

Paper. You must use specially manufactured duplicating paper, which is much more absorbent than ordinary paper. It comes in a wide variety of colours and a range of weights – from airmail paper, 47 g/m² (30 lb), 80 g/m² (55 lb) for single-sided work, and 80 or 90 g/m² (60 lb) for double-sided work, to board (180 g/m²) (120 lb) for tickets, covers etc. You can also buy postcards and gummed squares in sheets for duplicating. Paper is also made from recycled paper – it does not give quite such good-quality prints but it is very good value, costing about half as much as normal duplicating paper.

SPECIAL OPERATIONS

Printing on both sides of the paper. If you intend to print on both sides it is better to use 80 g/m² (55 lb) or 90 g/m² (60 lb) paper or card. If you find that the image is coming through the paper even though you are using a heavier-weight paper and you are not over-inking, you may have to print on single sides only, or on to card.

If you are rerunning the paper through the machine make sure that it is in a neatly stacked pile, because the printing on the first side may have buckled the paper so that it will not feed properly into the machine. The heavier the print the more buckled the paper will be and the higher the stack the more accentuated the buckling. Run smaller amounts at a time, and always print the side with fewest heavy solids first to minimize the problem.

Changing the colour. There are different methods for the two types of machine:
- *On a two-cylinder machine.* To change the colour of ink in the machine you use a 'colour kit'. This is a substitute set of all the parts that get inky, supplied in a storage tray. Changing the parts and cleaning the cylinders takes about five minutes. You need a colour kit for each colour, and they cost about £100 ($60) each.

OPERATING A STENCIL DUPLICATOR

Remove the screen, ink rollers, ink pump and distributor and tube of ink from the machine and store them in their tray.

Wipe all traces of the previous ink from the drums with a rag or paper towel.

Put in the parts from the new colour kit, making sure that the ink rollers are moving in and out correctly before replacing the screen.

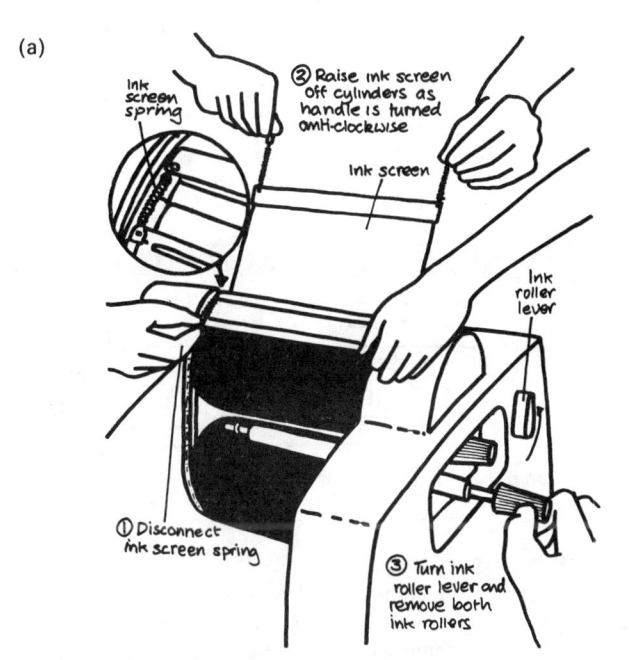

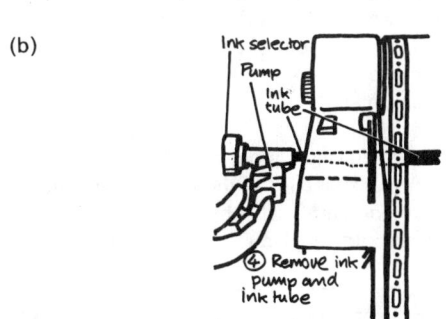

(a)

(b)

*Changing the ink in a two-cylinder machine:
(a) changing the parts; (b) changing the ink tube*

● *On a one-drum-type machine.* It is easier to change the ink in these machines than in the Gestetner type. The drum of ink is simply unlocked and taken out, and then the drum with the next colour is put in the correct position and locked in – you don't have to do any cleaning.

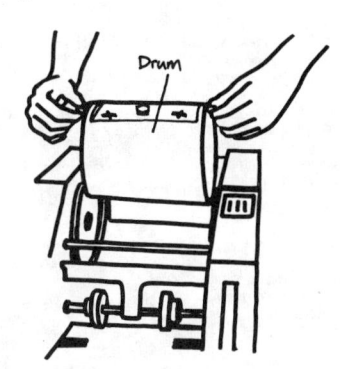

Changing the drum on a one-drum-type machine

Printing in two or more colours. Each colour in the design needs a separate stencil (⇨ page 18) and each stencil is used to print the appropriate colour. Always print the colour with the fewest heavy solids first, to prevent the paper buckling, as this may cause trouble when you come to print the second colour (⇨ page 18).

1. Put on the stencil for the first colour, print it as usual, then remove the stencil.

2. Change the colour of ink in the machine (see above).

3. Put on the stencil for the second colour and run scrap paper through until it is printing evenly.

4. Now start to print on the sheets printed with the first colour. After making a few prints stop and get the second colour in the correct position (in register) with the first colour by adjusting the print-height knob, and/or moving the paper pile to the left or right as necessary (⇨ page 24).

5. Print off the second colour.

6. Repeat steps **3** and **4** for subsequent colours.

CLEANING AND MAINTENANCE

Always clean the machine each time after use, with paper towels or rags. Clean off superficial ink smudges, ink that has collected at the sides behind the drive bands and the screen, and under the machine. Clean the pressure roller, the forks, the paper-weight stops, and anywhere else where there should not be ink. Do not wipe the screen itself because (*a*) bits of paper towel come off

Two-colour leaflet using electronically cut stencil (A4)

and clog it up and (b) you can easily damage it if you are not careful. The best way to clean the screen is to run through scrap paper until all the surplus ink is absorbed. When the machine is not in use always put a cover on it to keep out dust. These precautions will ensure that the machine remains in good working order.

Putting in new parts. It is worth stocking the following spare parts, so that you can replace parts yourself if they become worn or break:
- Screen.
- Ink rollers.
- Ink distributor and pump unit.
- Pressure roller.
- Drive bands/split pins.
- Forks.
- Rubber pads for paper feed.

When should you call in the mechanic? Call in the mechanic when you are not sure what is wrong, or whenever anything goes wrong with internal mechanisms. Watch the mechanic at work so that next time you may be able to do the same repair yourself and thus save time and money. Once you get to know your machine you will find that you can fix 80 per cent of what goes wrong.

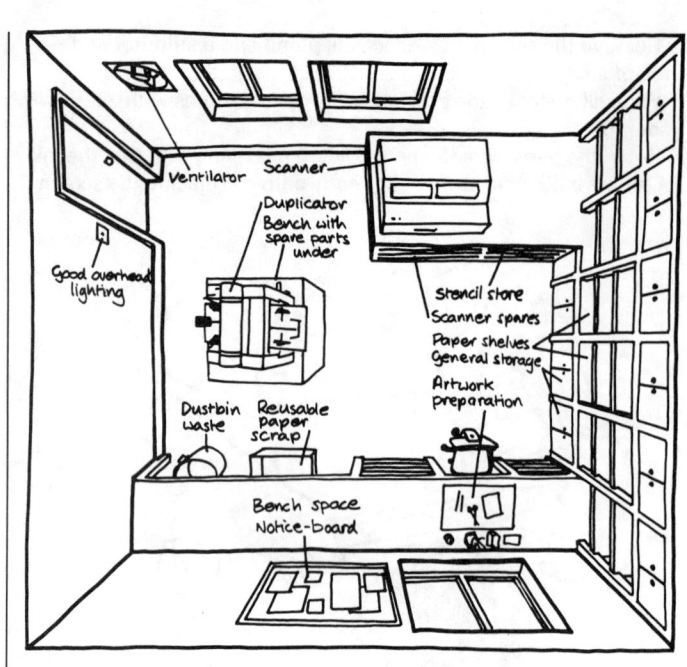

A possible arrangement for the print-room (bird's-eye view)

SETTING UP AND RUNNING A DUPLICATING PRINT-ROOM

One example of a successful duplicating print-room is the one run since 1974 by the Islington Bus Company, in London, for the use of local community groups. As part of a large resource centre funded by the local authority, the print-room is in almost constant use, and in the day-to-day running of the print-room the Bus Company have come up against and had to solve many problems. In this section some of their experiences are described. We hope that this account will help to encourage other organizations to open up their resources to local groups.

ARRANGING THE ROOM

If you are considering opening a print-room for the use of many groups, it will not be enough just to buy a duplicator and begin. Duplicating does sometimes start in this way but if there is to be much use of the facilities by a variety of different groups then thought needs to be put into the layout of the print-room. Perfect conditions are rarely obtainable. Nevertheless, it is worth knowing what to aim for as the ideal set-up.

The room. The first thing to do is to choose a light, well-ventilated room with plenty of space for all the equipment needed. It is also important to have good artificial light for use outside daylight hours. The room should be adequately sound-proofed; it

Hand-out to advertise the resources available at this resource centre (electronically cut stencil, A4)

is very difficult to print in the same space used for other activities – and the noise becomes very distracting and annoying to the other people. So if a print-room is to be an extension of your present activities, try to set it up in a separate room. It is also essential to lay down a no-smoking rule, because of the amount of inflammable material that there is in a print-room.

Basic equipment. The next thing to consider is the basic equipment needed. First, there is the duplicator, which should be simple to operate and easy to service. A scanner, although a relative luxury because of its high price, is well worth getting if at all possible, perhaps by means of several groups pooling their money. As with the duplicator, it is advisable to get an uncomplicated machine since it will be used by many different people with widely varying levels of skill.

Both the duplicator and the scanner vibrate a good deal and should stand on strong tables. The custom-built units made by the manufacturers can often be picked up second-hand from office-furniture shops or through classified ads, and they are also very useful for storing spares. If the overall lighting is insufficient, directional lighting (especially over the scanner) will be necessary. There should be free walking space around the duplicator, and plenty of general surface space for people to lay out their work, both of which should be free from clutter and cleared up by the groups after use. There should also be a table, shelf or drawer for all the additional odds and ends which people need in the print-room (corrector fluid, scissors, masking tape etc.). Provide a couple of boxes, one for test stencils and one for people to leave you a few copies of what they have printed. In a community print-shop, this material not only will be of general interest to yourselves and others but may also give other people ideas about layout, working methods and the like. You will need two large rubbish bins, one for rubbish and one for scrap paper which can be recycled. (Contact your local Friends of the Earth, or a waste-paper merchant, to arrange for collection of waste to be recycled.)

Other physical items which could be considered are: a light-box on which to check stencils; a 'jogger' to restack paper evenly into neat, straight piles for two-sided printing; a guillotine or paper cutter to cut paper and card to the required sizes for tickets etc.; and a collator – not a necessity, but when the print-room is much used it enables groups to save time in collating reports or newsletters of several pages.

Good use should be made of available wall space for posters and information, from simple instructions on how the machines work to examples of various colours of ink and paper available, and so on.

Layout space. Another facility to consider when setting up a print-room is whether to offer space for preparation of artwork for scanned stencils, or for typing out stencils. This could be a separate room, well lit with good work surfaces. It should include a typewriter (electric if possible, and ideally with interchangeable typefaces, for example a golf-ball machine) and a variety of additional art materials for preparing the paste-up ('Design and Paste-up' ⇨ page 85).

TRAINING

A fundamental decision to make before you start a print-room is about who is going to use it. Are you going to restrict the use of the equipment to people who know how to use it, or are you willing and do you have the time to train people – and if so, at what times? Showing people how to duplicate is a straightforward task but it requires patience and a uniform approach. It allows many more people to use the equipment and should also help to prolong the life of the equipment (since it is likely to reduce the amount of abuse of the machinery).

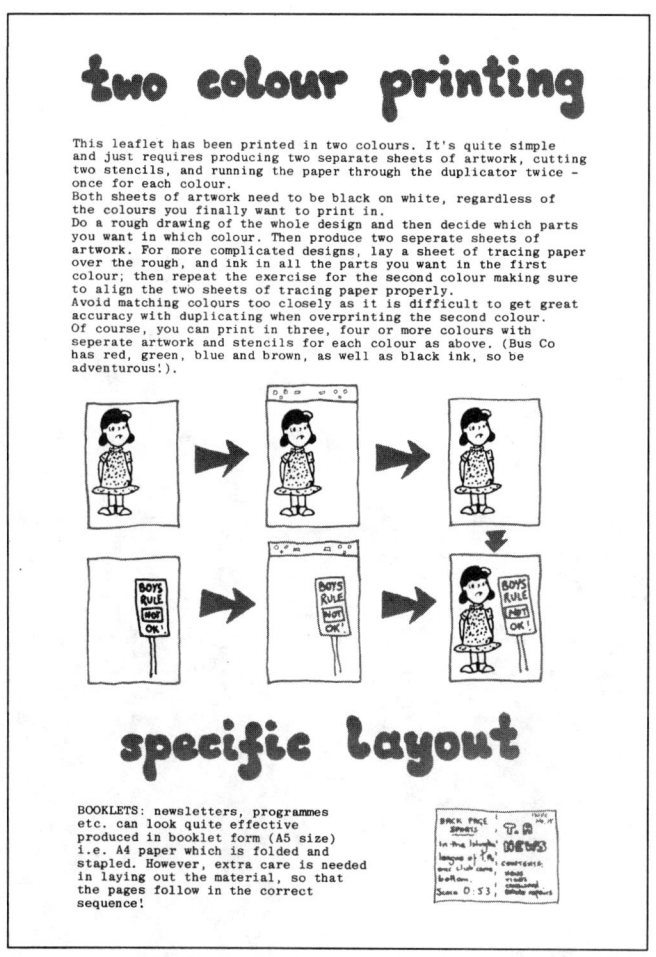

Page from Print It *leaflet produced by the Islington Bus Company to explain stencil duplicating (A4, electronically cut stencil, two colours)*

When people come in (or phone) you can save a lot of time and trouble by having clear information ready about costs and ways of preparing artwork and stencils (especially if this is being prepared beforehand at home). Try to have a uniform method of training so that people are not confused if they learn from different people on different occasions. Don't expect people to have learnt how to use the equipment after just one go; some pick it up straight away but others need much more training. And some people, however competent, may not print very often, and may forget how to use the equipment. It is better to be safe than sorry, so if people are not sure of the procedure, stay with them, for things can very easily go wrong.

The training should be thorough: explain the machine controls, the printing process, the technique of printing, the things that can go wrong and how to correct them and, just as important, how to clean the machine and clear up after use. Also, make sure that people using the machinery call for help as soon as they need it, without trying to fix things themselves when they are not certain how to.

DAY-TO-DAY RUNNING

If there is much use made of the facilities, you will need a booking system. Always overestimate the time needed: things often take longer than you expect. A rule of thumb is that untrained people take fifteen minutes to scan one sheet and forty-five minutes to print 500 sheets; a trained person also takes fifteen minutes to scan one sheet but only five to ten minutes to print 500. (The times obviously vary considerably with the types of machine and the stencils you use.)

A typical layout for a duplicating-room (the Islington Bus Company)

The decision about who you will allow to use your duplicating-room is yours, and if you are going to run the room satisfactorily it is a decision that you should make carefully. Do you want profit-making organizations to use the facilities? How many people at a time? When can you train people? Who is going to do the training? A print-room (like many other community resources) is quickly given up by people if it is unreliable. Many people who will be using the facilities have to make time to do so in a very busy day, so if your machinery breaks down make sure that you can contact them and warn them before they come.

The print-room users will be as varied as any other cross-section of people. Some will need a lot of coaxing, others will take to the new technique immediately. You will need patience. Be careful not to overload yourself with work. Try not to cram in too many people in one day, and try to leave some time for the 'must have this newsletter tonight' group. Always overestimate the time needed to do a duplicating job.

The rewards for all this are enormous. First, there is the genuine pleasure you can get from teaching someone a new skill. Then there is the knowledge that you are helping the people in your locality to get their views and news across. You will also find that as various groups pass through the print-room more and more insight is gained into your overall knowledge of the area. Talk to people while they are there and soon you will find that the print-shop becomes an information centre and a news swap shop. People will call in to find out what is going on as well as to print, and thus your duplicating-room becomes a valuable local resource.

MANAGING THE PRINT-ROOM

Repair. You will have to arrange an efficient maintenance service on all your machines. The manufacturers all have maintenance facilities, but it may be better to get to know your local mechanics as they are usually cheaper and can teach you simple maintenance, which will save you time and money. When you do have to call in the mechanics be sure to watch them, ask questions, and learn how you can fix the problem yourself.

Supplies. Make sure you have an adequate supply of materials. It is difficult to forecast demand when starting, but you should order well in advance of actually needing the materials. If you have the room and can afford it, it is always better to overstock, because prices inevitably go up and you can take advantage of special bargains.

Costs. If you are running a non-profit-making print-room, work out your price so that you cover the cost of materials, and allow a small margin to cover the inevitable wastage that will occur of stencils, ink and paper. Groups which do not get funding that covers their running costs may also have to work out an extra charge to cover overheads like heating, lighting and electricity.

If the funding of your project does not cover the salaries of the workers who are going to run the print-room and train people, you will have to pass this cost on to the users. (The Islington Bus Company, like many other resource centres in the U.K., relies on local-government funding to pay its workers.)

THE FLAT-BED DUPLICATOR

As the name suggests, the flat-bed duplicator is a flat stencil duplicator, like a small screen-printing frame. It uses normal duplicator stencils, ink and paper, but it is worked completely by hand. There are many different designs but the general principle has been in use for a long time. The Victorian duplicator was a flat-bed housed in a beautiful wooden box that could be carried around. The frame and baseboard are made of wood, the screen is made of fine-mesh nylon, and an iron rod is used to press the ink through the stencil on to the paper. The frame and screen are pivoted to supports a few inches above the baseboard, so that paper can be placed under the screen.

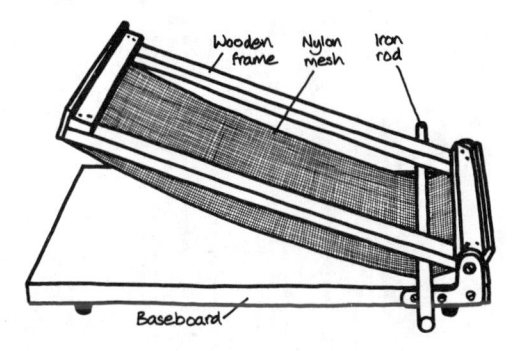

A flat-bed duplicator

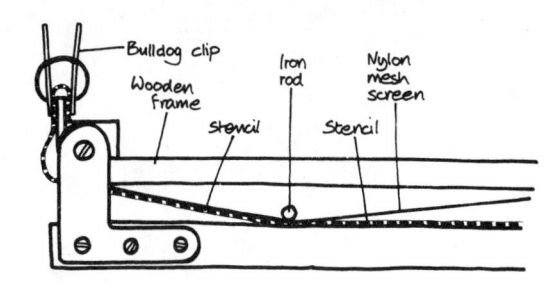

How the flat-bed duplicator works

A hand-cut stencil-duplicated newsletter (A4)

Using a flat-bed duplicator

1. Squeeze a ribbon of ink across the top of the mesh, and with the screen raised spread the ink by pulling the rod down the mesh. Leave the rod at the far end.

2. Attach the stencil with small bulldog clips or spring clips to one end of the frame, making sure it is face up and can be read as you look at it through the screen. Make the stencil adhere to the screen by gently rolling the ink rod down the screen, ensuring that the stencil does not wrinkle.

3. Place a small pile of paper under the stencil and make the first print by pulling the rod from the top to the bottom of the screen. Lift the frame up and remove the print. Check that it has been printed in the correct position on the paper and if necessary move the pile of paper to a better position. Lower the frame and continue printing. When the mesh starts to run out of ink squeeze more into the top of the screen.

Paris, 1968, poster

Screen Printing

Introduction 33
How does screen printing work? 33
When would you use the process? 33
Relative costs 34

Equipment: what you need and how to make it 34
The frame 35
The mesh 36
The squeegee 37
The surface to print on 38
Drying systems 38
Preparing the screen 40

Simple stencils 41
Cut paper 41
Liquid filler 42
Reversal with filler 44
Cut film 44

Making photo-stencils 46
The positive (hand-drawn, photographic, and for multicolour work) 46
The light source 49
Producing the photo-stencil 50

Making the prints 52
Inks 52
Paper 52
Planning your printing area 52
Preparation and registration 53
Taking a print 53
Printing more than one colour 54
Cleaning up 55

Other useful information 55
Printing T-shirts 55
Larger equipment 58
Finding and using a commercial printer 59
Specialist suppliers 59

INTRODUCTION

This chapter explains what the screen-printing process is, and how you can build and use all the simple equipment that you need to make prints in this way.

HOW DOES SCREEN PRINTING WORK?
Screen printing is basically a form of stencil printing:

1. A wooden frame is stretched tightly with a fine fabric (the mesh) and this is called the screen.

2. A stencil is put on to the mesh (in one of many ways) and this blocks some of the mesh and leaves some of it as open areas.

3. The screen is laid down on top of a sheet of paper, and a rubber blade (called the 'squeegee') pulls the ink across the inside of the screen.

4. Ink is pushed through the open areas of the stencil on to the paper below to make a print.

5. The screen is lifted and the paper removed and hung up to dry. The process is repeated until the number of prints needed has been printed.

WHEN WOULD YOU USE THE PROCESS?
Screen printing is one of the most useful printing methods for someone wanting to make prints cheaply, because the equipment can be built and used quite simply and the resulting prints are so bold and clear. The most suitable occasions to use screen printing are when you need to print:
● a run of fewer than 200 copies.
● designs with large solid areas of colour, but not very fine details.
● prints larger than A3 ($16\frac{1}{2} \times 11\frac{3}{4}$ in.) size.
● on to fabric, T-shirts, wood, glass, plastic or metal.
● and if you've got the time to do it yourself!

Printing posters in the screen-printing workshop

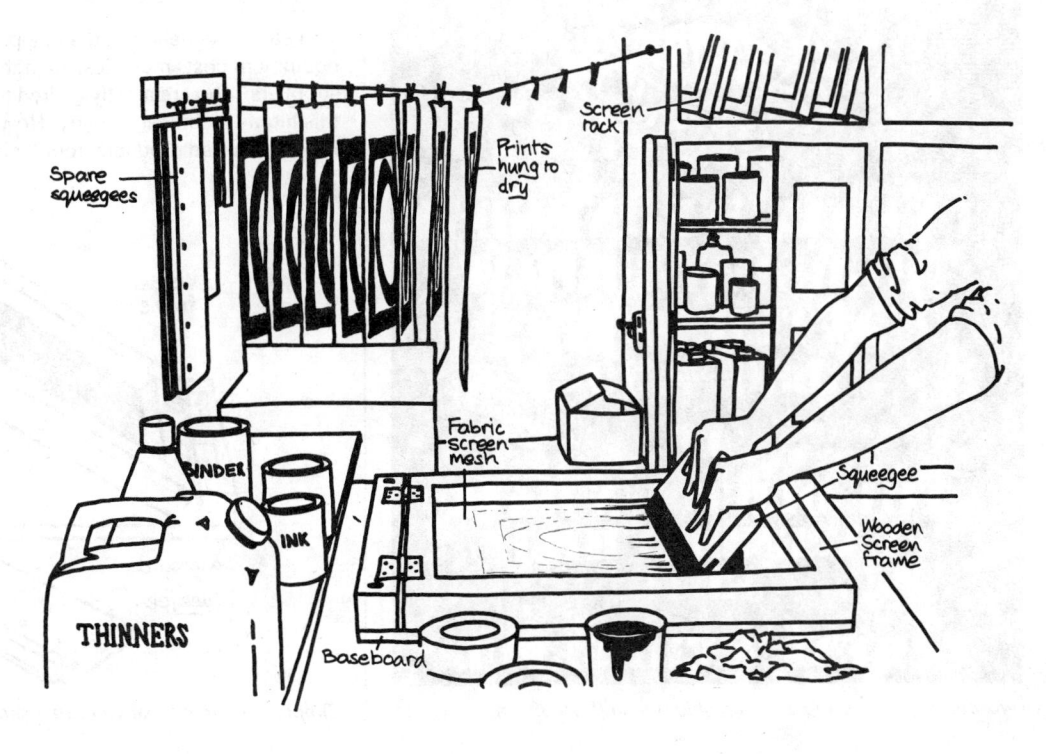

Page from a women's calendar, painted-filler stencil (50 × 39 cm)

Two-colour poster using photographic stencil (50 × 39 cm)

Lettering for a one-colour poster using cut-paper stencil

RELATIVE COSTS

Once you have covered the cost of building your equipment, screen printing your own posters should reduce the expense to one-half to one-third of what it would cost to use a commercial screen printer. The cost is affected by:

- *Number of copies:* this will determine the amount of ink and paper you use.
- *Number of colours:* in most cases you need to make a separate stencil for each colour on your poster (*but* ⇨ page 54 for ways round this!).
- *Type of stencil:* because the cost of the stencil is the same however many prints you make, bear in mind the length of the run when you choose your stencil method; some methods are very expensive and the cost may not be justified for a very small number of posters.

EQUIPMENT: WHAT YOU NEED AND HOW TO MAKE IT

You can make nearly all the equipment you need. Home-made equipment costs much less than that manufactured by specialist suppliers, who are mostly geared to dealing with big business in this highly technical industry. However, you may need to buy a few specialized products from their suppliers.

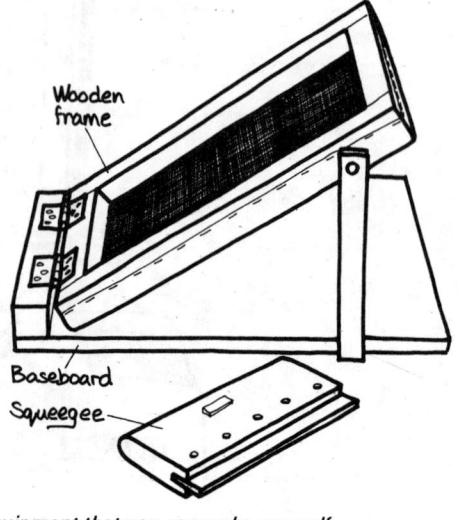

Simple equipment that you can make yourself

EQUIPMENT: WHAT YOU NEED AND HOW TO MAKE IT

If you intend to print long runs (into the few hundreds) or complex multicolour prints, and you think you will be doing so regularly, it might be worth buying or making some larger pieces of equipment, such as a drying-rack or vacuum table (⇨ page 58). But you will be surprised by how much printing you can do with only the basic home-made equipment.

THE FRAME
You need a sturdy, rigid frame with strong corner joints that will not warp; the frame must lie flat when placed on the printing surface. Use wood that is straight and knot-free; cedar is light and strong but expensive; deal is cheaper and easy to get but not really of such good quality.

The size and shape of the frame depends on what you need to print. It is wise to build frames to take the size of poster that you will most often be printing; you can always print a smaller poster on a large screen but not the other way round! Whatever size you choose make sure you allow at least 20 cm (8 in.) more on the width, and 30 cm (12 in.) more on the height; this is to give you plenty of room around the print area for the ink to collect. Do not skimp on this or it will cause endless problems when you are printing.

The thickness of the timber you should use for the frame depends on the size of the frame itself: frames up to 75 × 50 cm (30 × 20 in.) in size require 5 × 5 cm (2 × 2 in.) timber, but larger frames will need to be made with 7·5 × 5 cm (3 × 2 in.) timber.

Here's one simple way to build the frame. This frame is for printing posters in sizes up to 51 × 38 cm (20 × 15 in.) so it has an inside size of 81 × 58 cm (32 × 23 in.). You need:
- Six lengths of timber 63 × 5 × 1·5 cm (26 × 2 × ¾ in.).
- Six lengths of timber 86 × 5 × 1·5 cm (32 × 2 × ¾ in.).
- Sixteen 30-mm (1-in.) panel pins.
- Woodworking glue, brads, hammer, clamps, sandpaper.

1. Take three of the first six lengths of timber, and glue and nail them flat together with the middle piece offset by 5 cm (2 in.), making sure that the two outer pieces lie directly above each other.

2. Repeat this process to make up the three other sides so that you produce four pieces measuring 5 × 4·5 cm (2 × 1¾ in.) in section, two of one length (68 cm/27 in.) and two of another (91 cm/36 in.).

3. Position the four lengths so that the overlaps at the ends are ready to make joints (similar to tongue-and-groove joints).

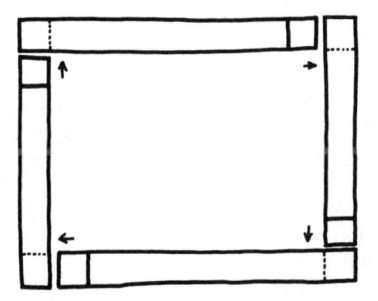

4. Glue the joints, then lay the frame on a flat surface and, having checked that the corners are square and that the frame lies flat, nail the joints with two panel pins/brads in each side of each corner.

5. Leave to dry overnight on a flat surface with a weight or clamps on each corner.

6. Sandpaper the outside edge of the bottom of the frame to round off the sharp edge – this is to make stretching the mesh easier.

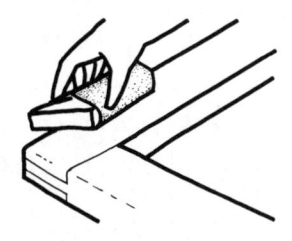

THE MESH

The mesh is stretched drum-tight across the frame. It holds the stencil in place, but allows the ink to pass through. Many fabrics can be used; it depends on how much you can afford to spend and on what quality of print you need to make. The cheapest and easiest meshes to find are the organdies:
- *Cotton organdie* is the cheapest but sags a lot after it has been used a few times; because it is not strong, it can be very difficult to stretch without tearing.
- *Terylene organdie* is easy to find – it is very fine net curtaining. It costs more than cotton but it is much stronger, can be stretched better and lasts longer, and is the ideal mesh for the beginner.
- *Specialist meshes* are available from specialist suppliers and come in an enormous range of types and qualities which can be very confusing. These meshes cost three or four times more than organdie but are much more durable and if looked after will have a very long life.

A finer mesh is a must if you need to print fine details, halftones or other high-quality work. Nylon is easier to work with than polyester (the two main types) because it is easier to clean out, but it needs to be 'degreased' every time a stencil is attached or the next stencil may not adhere well (⇨ page 40).

Mesh count. When buying specialist meshes you have to decide on the mesh count (how many threads to the centimetre) and weight. The weight is indicated by a letter: S is woven from thin threads, T is medium, and HD is extra heavy. The mesh count that you need depends on the type of work you are printing. Here is a rough guide:

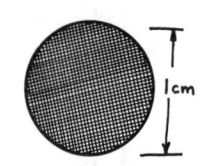

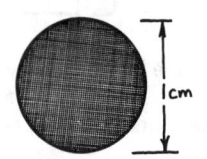

Close-up of two meshes, fifty and one hundred threads to the centimetre

- **43T** For fabric printing and for printing open areas of colour with hand-cut stencils.
- **77T** For general work and detail down to about the equivalent of 20-pt type (⇨ page 90).
- **100T** For fine work, such as halftones down to about 45 dots to the inch.

Note that the finer the mesh the more it costs – and the harder it is to keep clean – so do not use specialist mesh unless you really need to for high-quality work.

Stretching the mesh. Organdie can easily be stretched on to the frame by hand, but with specialist meshes it is difficult to get the correct tension and evenness and the frame is best stretched by machine. Most mesh suppliers have a stretching service. Phone them to see how much it would cost. The price is worth paying, since a machine-stretched mesh will wear better and last longer.

Stretching by hand

1. Cut a piece of mesh about 15 cm (6 in.) larger than the outside size of the frame. Wet it with water (it stretches more easily wet).

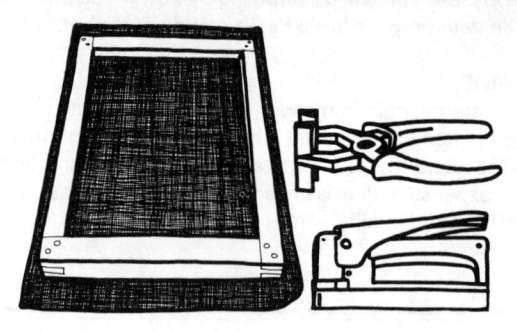

2. Lay the fabric in position across the frame, making sure that the threads are running at right angles to it.

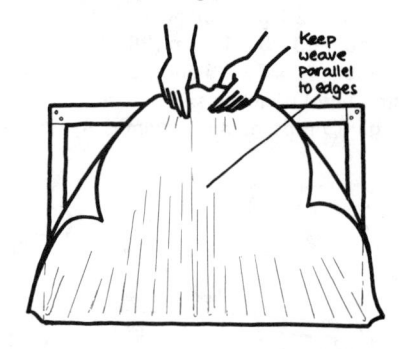

3. With a staple gun, staple the mesh to one of the short sides, starting at the middle and working out to the edges.

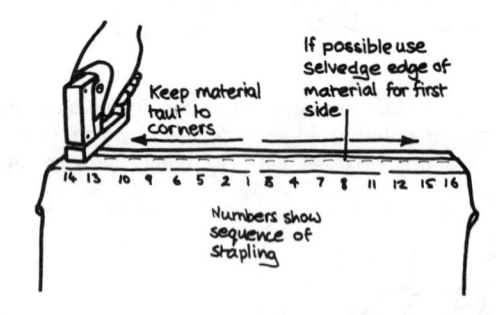

4. Staple the opposite edge in the same way, making sure that you pull outwards and sideways very tightly.

5. Repeat steps **3** and **4** on the two remaining sides to obtain a screen which is drum-tight with no puckers or unevenness in the mesh (staple down the overlap at the corners).

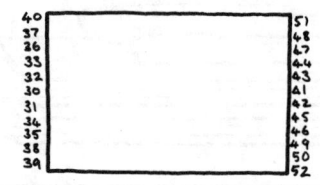

EQUIPMENT: WHAT YOU NEED AND HOW TO MAKE IT

THE SQUEEGEE
The squeegee is the tool used to pull the ink across the screen. As it passes over the open parts of the stencil it forces the ink through the mesh on to the paper below and makes the print. The squeegee consists of a straight-edged blade (usually rubber or polyurethane) sandwiched between pieces of wood which make the handle. A squeegee can be bought ready-made from suppliers but it costs less to buy just the rubber or polyurethane strip from them and make your own holder. If possible make it so that the blade can be removed and changed if it should get damaged.

Size. Your squeegee needs to be at least 5 cm (2 in.) wider than the area you will be printing; however, do not use a squeegee that fits too tightly just inside the screen as this will cause problems when you are printing. You will need a few squeegees of different sizes: make sure they link in with your screen sizes and common printing sizes.

Here is one way to make a squeegee. To make a 45 cm (18 in.) squeegee you need:
- Two lengths of planed timber 45 × 7·5 × 1·5 cm (18 × 3 × ¾ in.).
- One length of planed timber 45 × 5 × 1 cm (18 × 2 × ½ in.).
- One 45-cm (18-in.) strip of 1-cm (½-in.) blade (of either white rubber or soft polyurethane).
- Six bookbinding posts and screws 4 cm (1½ in.) long (available from stationers).
- Woodworking glue, Stanley knife or plane.
- Sandpaper, polyurethane varnish and brush.
- 1-cm (½-in.) drill and bit, screwdriver, clamps.

1. Lay the three pieces of timber flat on top of each other, sandwich-style (the narrower piece in the middle) with edges aligned, and glue in place and clamp from sides. Leave to dry.

2. To make the squeegee easier to hold, round off the corners of the back or top by planing and then smooth off with sandpaper.

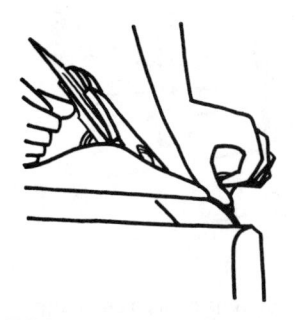

3. Give the whole thing a coat of polyurethane varnish and leave to dry.

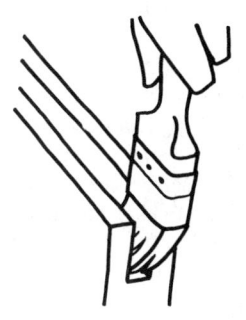

4. Place the blade in the groove of the holder. Clamp it to the edge of a bench and drill holes at 7-cm (2¾-in.) intervals through both wood and blade.

5. Put in the posts and screws and tighten them up.

6. Glue a small block of wood on to each side of the squeegee – this provides a ledge to support the squeegee to stop it falling into the ink between printings.

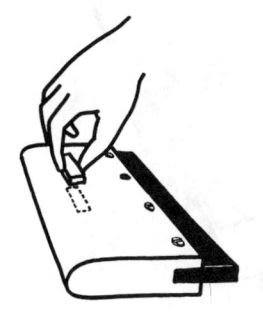

Care of squeegees. For good crisp prints the squeegee must be kept sharp and undamaged, so:
● Clean squeegees well after use.
● Store them hung up on a hook so that the blade does not get knocked and damaged.

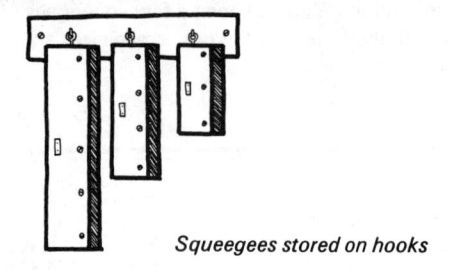

Squeegees stored on hooks

● Keep the blade sharp by using a squeegee sharpener. This can be made by sticking some coarse-grade emery cloth on to a board, attaching another length of timber along the side to act as a guide and to keep squeegee upright. The blade is sharpened on the board by being pulled gently backwards and forwards across the emery-cloth.

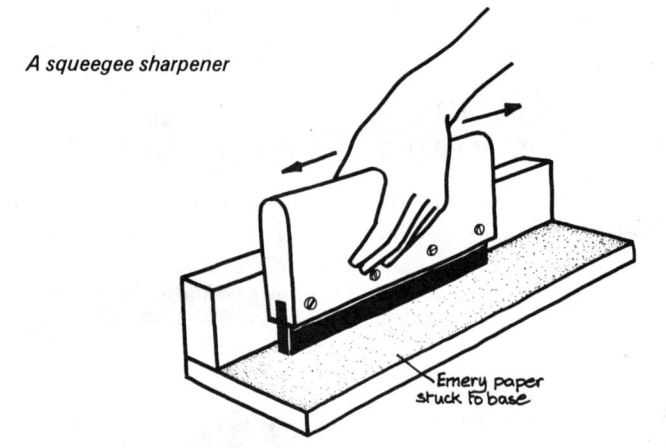

A squeegee sharpener

THE SURFACE TO PRINT ON
You need:
● A smooth, flat, easy-to-clean surface on which to place the paper.
● A way of hinging the screen to the surface so that it can be raised and lowered and always goes down into the same position on the surface for printing.

One way to produce your work surface is to attach a bar of wood to the edge of your printing table. This can be a temporary system using two G-clamps, or it can be permanently glued and screwed down. The screen is then hinged to the bar, and loose-pin hinges are used so that the screens can be changed and removed for cleaning.

Screen and bar clamped to printing table with G-clamps

One other very convenient method is to make a portable 'baseboard'. This enables you to print almost anywhere, and it also makes it easier for you to keep the printing surface in good condition, because it can be stored out of the way when not in use. (This is very handy if you have to use a room that is also needed for other activities.) One way to make a baseboard is shown opposite.

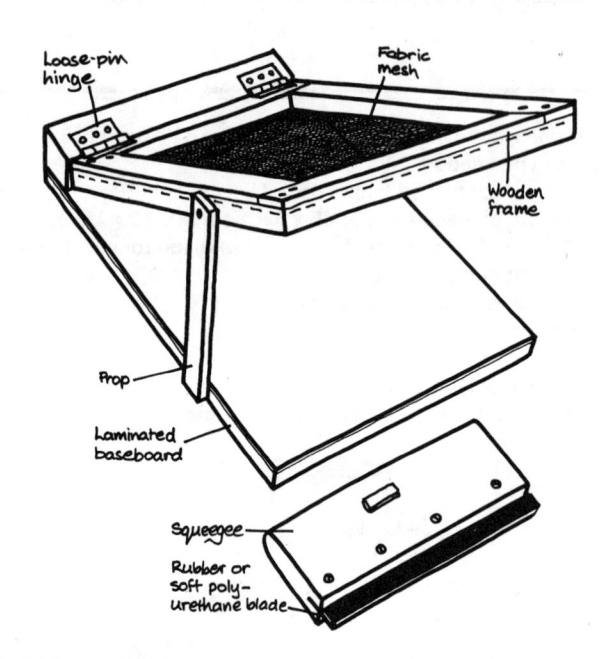

Portable baseboard

DRYING SYSTEMS
Most inks take at least twenty minutes to dry, so you need to be able to hang up or lay the prints somewhere (without their touching each other) until they are dry. Whichever method you use, try to do the drying near to your printing table so that you do not have to spend extra time walking back and forth carrying wet prints.

1. Hang the prints with pegs, paper-clips or bulldog (spring) clips from a washing line of cord or heavy string. (If large-sized prints curl up too much try hanging them from one corner only, or try method **2**.)

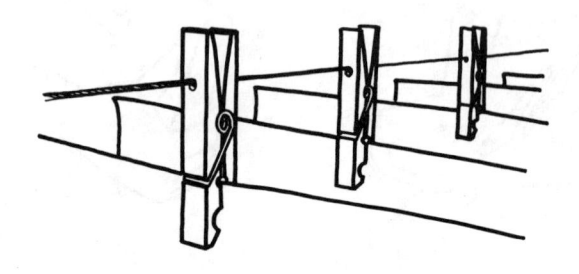

2. For larger-sized prints use two rows of pegs or clips and hang the prints from two corners.

EQUIPMENT: WHAT YOU NEED AND HOW TO MAKE IT 39

One way to make a baseboard. You need:
- One piece of planed timber 80 × 5 × 5 cm (32 × 2 × 2 in.).
- One piece of laminated chipboard or flakeboard 100 × 80 × 2 cm (40 × 32 × ¾ in.) and four 5-cm (2-in.) screws.
- Two 7·5-cm (3-in.) loose-pin hinges.
- Screws, screwdriver, clamps.

For the prop you need:
- One length of timber 25 × 3 × 1 cm (10 × 1¼ × ½ in.).
- One 2·5-cm (1-in.) screw, drill and bit.

1. Drill four holes through the base of the chipboard on one of the short sides.

2. Clamp the bar in position along the same short edge of the laminated side of the board, and put in the 5-cm (2-in.) screws through the holes you have drilled.

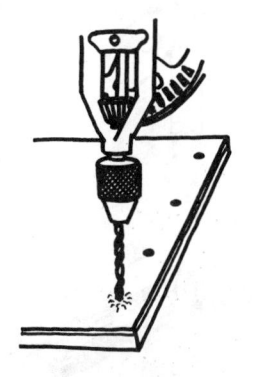

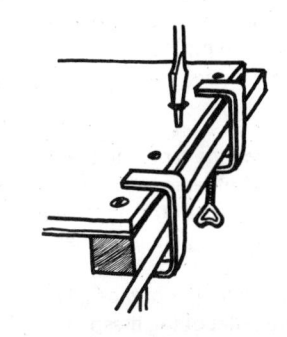

3. Remove the pin from one of the hinges. Attach the two parts to the hinge bar, one at each end.

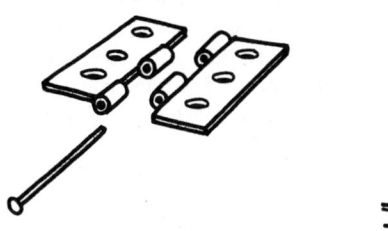

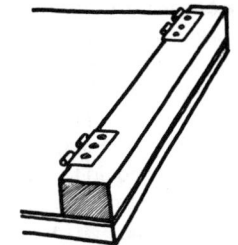

4. Lay the screen in position with a short edge up to the hinge bar. Take the second hinge apart and match it up to the two parts already screwed to the hinge bar; put the pins back in and screw the remaining sides of the hinges to the frame. (This must be done very securely so that the frame cannot move from side to side.)

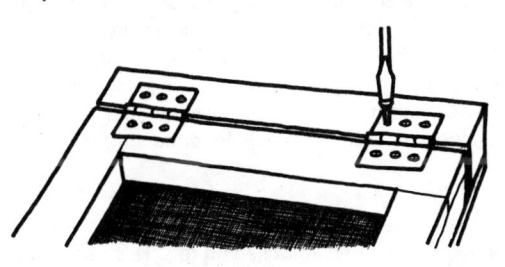

5. To make the prop, which will hold the screen in a raised position between printings, drill a hole in one end of the piece of timber and screw it to one side of the frame using a screw slightly smaller than the drilled hole. This allows the prop to swivel and to drop down as the screen is lifted, which it supports in the raised position.

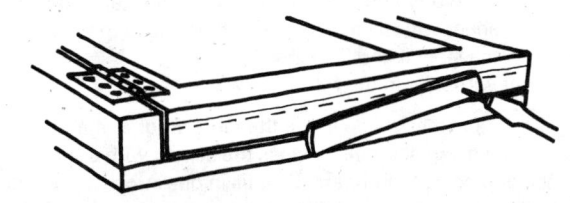

---

3. A 'ball and pattern' drying-rack, in which a small marble grips the paper, is available commercially.

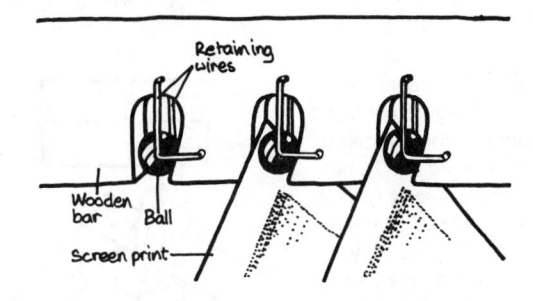

4. Try to build your drying-rack so that once full it can be hoisted up out of the way by some sort of pulley.

5. Commercial drying-racks (⇨ page 58).

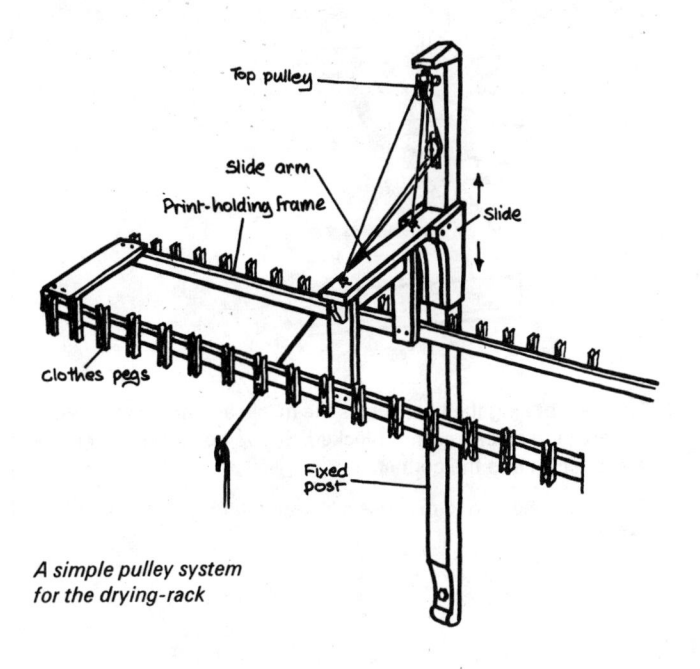

A simple pulley system for the drying-rack

SCREEN PRINTING

PREPARING THE SCREEN
Depending on which type of stencil you are using, various preparations may need to be made to the screen.

Degreasing the screen. A newly stretched screen must be degreased before use. To do this stand the screen in the sink, wet it and pour on some washing-up liquid or degreasing agent (such as Pregan Paste), from specialist suppliers. Scrub the mesh lightly from both sides with a scrubbing-brush or washing-up brush. Leave it for a few minutes, then rinse off well with clean water and leave it propped up to dry. Screens always need to be degreased if you are going to use water-adhering or presensitized photo-stencil films, because any grease in the mesh will prevent the stencil from adhering properly.

Taping up the screen. To prevent the ink from seeping out under the screen when printing, brown gummed tape should be stuck into the corners where the mesh meets the frame. If you are using cut-paper, liquid-filler or coated-screen stencils this job is best done before applying the stencil. But with other types of stencils (cut film, presensitized film) it saves time to adhere the stencil first and then to tape up, so that both stencil and tape can be drying at the same time. Some of these stencils have to be adhered to a wet screen, in which case it would not be practical to stick the tape first, and this should be done after the stencil has adhered.

Screen filler. The space left between the adhered stencil and the gummed tape needs to be blocked up too. This can be done by continuing the gummed tape right up to the edge of the stencil, but it is better to use screen filler since this is transparent and enables you to see through the mesh, which makes registration easier; it also has a smooth surface across which it is easy to pull the squeegee.

If you are using a stencil that has a backing sheet, apply the screen filler before removing the backing sheet; in this way you will not have to worry about going over on to the stencil because the backing sheet will still be protecting it. Remove the backing sheet when the stencil and filler are dry.

Block out with filler in the following way:

1. Lay the screen mesh-side up on the table. Pour a small pool of filler on to the mesh and pull it across the screen with a small squeegee or piece of cardboard. Do this in one stroke if possible so that you get a thin layer of filler blocking the mesh.

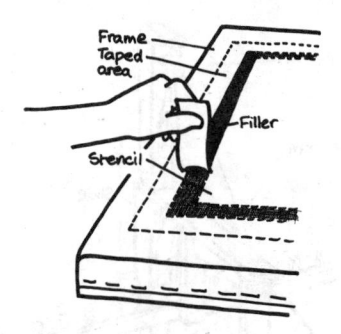

2. Continue pulling the filler across the mesh, adding more when necessary until all the mesh is blocked. Scoop up any left-over filler and return it to the container.

3. Prop the screen up to dry (use a hot-air fan to speed this up if necessary).

To tape up the screen. You will need:
- a roll of gummed tape (at least 5 cm/2 in. wide).
- a bowl of water.

1. Lay your screen mesh down on the table and raise it at one end so that the table is not in contact with it. Cut eight lengths of gummed tape to fit the inside edge of the screen; four to fit the length and four to fit the width. With the gummed side down, make a crease along the middle of the tape by folding it in half all along its length. Do this to two pieces of tape of each length.

2. Dip the tape right into the water, lift it out and get rid of the excess water by holding the tape between two fingers and running them down its length.

3. Stick the tape into the right angle between the mesh and the frame, stretching it slightly as it is being stuck so that it does not pucker. Stick a second length of tape all round on the flat of the mesh, overlapping on to the first piece. Stick two more strips of tape across the top and bottom of the screen to make the well into which the ink is poured and where it sits between printings.

4. Reinforce the corners with small pieces of tape, folded and cut as shown. If you are going to use water-based inks, paint the tape with shellac or varnish to stop the water dissolving it. (An easier and quicker way is to use masking tape instead of gummed tape, but this is more expensive and not as reliable.)

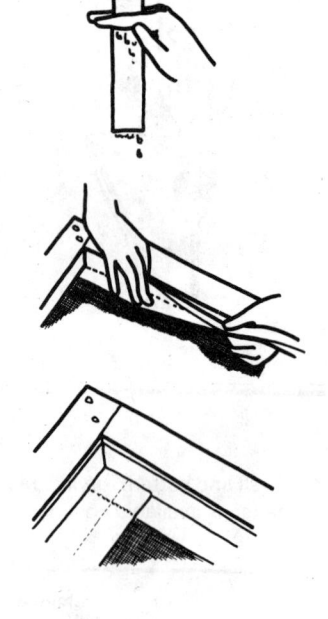

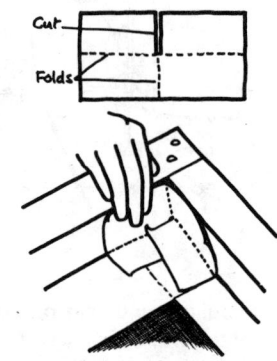

SIMPLE STENCILS

Anything that blocks the mesh or in some way stops the ink getting through the mesh can be used to make a stencil. The areas that remain unblocked, or open, allow the ink to pass through and this makes the printed image. In this section we will describe stencils made with cut paper, liquid filler, reversed fillers and cut films. The next section will go on to describe the more complex photo-stencils.

CUT PAPER

This is an ideal method for printing T-shirts or short runs (up to fifty copies) of simple posters. It is cheap and quick and can be used with either water- or oil-based inks (although the stencil lasts longer with oil-based inks). A thin paper is best for the stencil: newsprint, tracing paper and layout paper are all cheap and easy to get hold of.

Simple shapes are the most suitable to cut out by this method. If you are using words, try to design them without loose middles. If they do have 'floating' pieces, such as the middles of O's, keep these and put them back in position when you are ready to print.

T-shirt printed with cut-paper stencil

Above: One-colour poster using cut-paper stencil – note how the middles of the small letters have been left out (50 × 40 cm)

Right: One-colour poster using cut-paper stencil shows the detail that can be achieved with careful cutting (60 × 42 cm)

42 SCREEN PRINTING

Making a paper stencil. You need:
- a sheet of paper about the same size as your screen.
- a sharp scalpel or art knife (like an X-Acto).

1. Draw your design on the middle of the sheet of paper. Cut out with the knife the areas that you want to print; you will then have a sheet of paper with holes in it, which is your stencil.

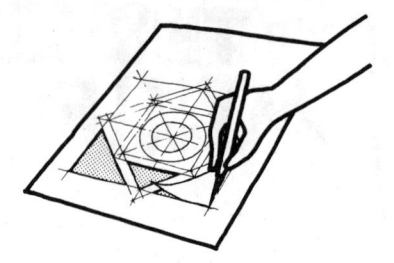

2. The stencil is now laid under the screen and with the first pull of the squeegee the wetness of the ink makes the paper stick to the screen (⇨ page 52).

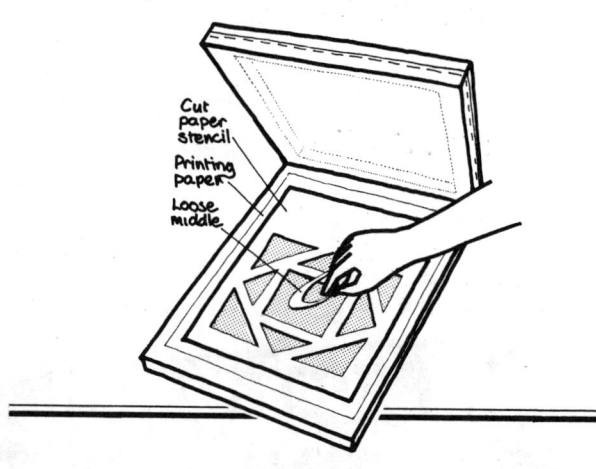

Paper stencils can also be made from 'found' objects such as doilies, computer tape, lace, Chinese paper cut-outs, etc. These can be laid under the screen and used in the same way as a cut-paper stencil. Some things can actually be stuck on to the underside of the screen:
- *Tape:* Sellotape or Scotch tape, masking tape.
- *Adhesive plastic:* Fablon, for example. Cut it out in the same way as paper, but stick it to the underside of the screen.
- *Sticky labels:* Labels come in many shapes and sizes and can be used to make borders, patterns and even words.

These stencils can only be used with water-based inks, since the solvents in oil-based inks tend to dissolve the glue in adhesive tape and plastic. If you are using oil-based inks try working with materials that stick on with water, like brown gummed tape or gummed-paper shapes.

LIQUID FILLER

A liquid filler is something that can be painted straight on to the screen to block up the mesh and make a stencil. What you use as a filler must be something that will not be dissolved by the type of ink employed, and that can be removed from the mesh when you have finished printing; thus you need a water-based filler when printing with oil-based inks, and vice versa.

Water-based fillers to use with oil-based inks include:
- Water-based glues such as School Glue and Le Page's Mucilage: Make sure you use a glue that will wash out again with water once it has dried.

Chinese paper cut-outs used as a paper stencil

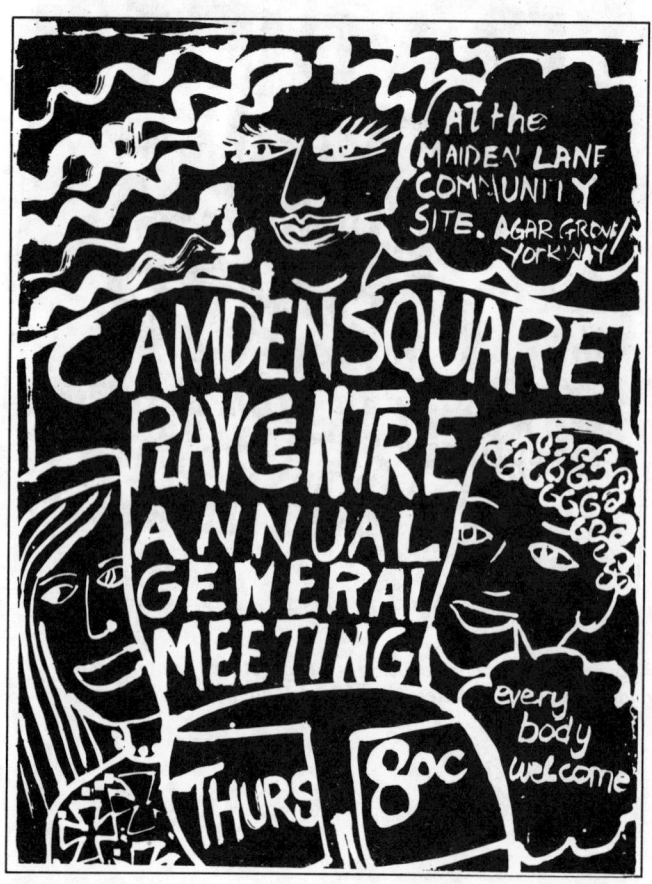

One-colour poster to advertise playcentre meeting using painted-filler stencil (58 × 47 cm)

SIMPLE STENCILS 43

● Screen filler (from specialist suppliers): Use a slow-drying type – it would be impossible to use a fast-drying type (which is also available) in this way.
● Gum arabic: Soak the crystals overnight in water, then bring to the boil, stirring the solution until all the crystals have dissolved. Leave to cool and store in an airtight container.
● Water-based paint such as poster paint or gouache.

Fillers and solvents for use with water-based inks

Filler	Solvent
Oil- or cellulose-based printing ink	White spirit or mineral spirits, screen wash
Liquid litho ink ('tusche')	White spirit, or mineral spirits, screen wash
Litho crayon	White spirit or mineral spirits, screen wash
Nail varnish/polish	Nail-varnish remover
Wax crayon	Paraffin
Candle wax (used as a crayon or melted and painted on)	Paraffin

Remember, the area that prints is the open area . . . the bit you did not paint out. So before you start, decide whether you want the shape or the background to be the white area. The easiest thing is to block out the words and print the background area, but it is possible the other way round, it just involves more work!

Do you want the shape – or the background – to be white?

How to make a filler stencil. You need:
● a roll of gummed tape.
● liquid filler.
● brushes.
● a soft pencil.

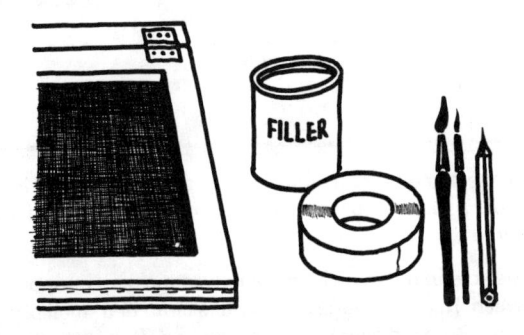

1. Before you start painting the stencil, put brown gummed tape into the corner of the screen where the mesh meets the sides of the frame. (If you are using water-based inks use masking tape instead). At this stage you can also block up the rest of the screen with brown tape or screen filler and leave just a 'window' open for your design to fit inside (⇨ page 54).

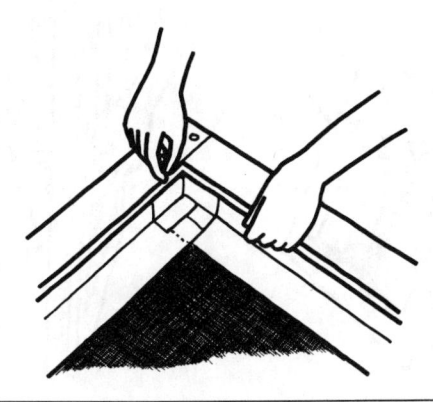

2. Draw your design the size you want it to print on a sheet of paper. Lay the screen face down on top of it and trace the design lightly on to the mesh with a soft pencil.

3. With the screen propped up so that the mesh can't touch the table and get smudged, paint the filler on to the mesh in the areas that are to stay white.

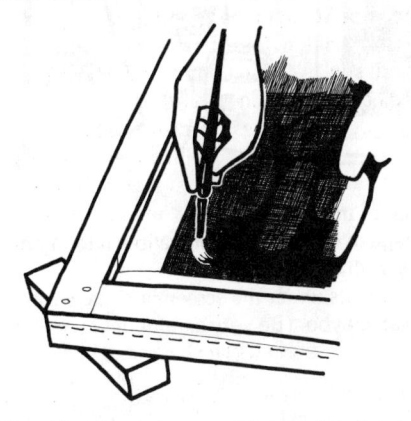

4. Let the filler dry with the screen lying horizontal, to avoid dribbles running from the wet filler. (A fan-heater helps to speed this up.)

5. Hold the screen up to the light to check that there are no little holes in the filler. Paint out any you see with some filler and allow to dry. The screen is then ready to print.

REVERSAL WITH FILLER

Once the design has been painted on to the screen with filler, it can be 'reversed out' so that the area that prints is the part you first painted, instead of the background.

Reversal with filler can create interesting textures

To do this you need two fillers with different solvents (⇨ list on page 43). If you are printing with oil-based inks the image is first painted on to the screen with an oil-based filler. When that has dried the screen is coated all over with a water-based filler. (Because oil and water do not mix, the oily image repels the water-based filler and it fails to stick on those areas.) When this second coating is dry, the first (oil-based) image is removed with screen wash, and this leaves open areas in the coating which make the stencil. Depending on the kind of ink you will be printing with, this process can be done the other way round, but make sure that, whatever you do, you end up with the stencil made out of a filler opposite in base to the printing ink you will be using.

CUT FILM

A cut-film stencil is much stronger than a paper stencil and can be used to make hundreds of prints. It should also be used for intricate designs that need more detailed cutting out, since the film is attached to a transparent backing sheet which holds all the pieces in place until the stencil has been stuck on the screen.

The different brands that are available from specialist suppliers vary according to the type of ink that is used. Some brands (called 'water-adhering') stick to the screen with water and are made for use with oil-based inks; other brands (called 'lacquer-adhering') stick on with a lacquer or other solvent and are for use with water-based inks or dyes. Each brand needs to be used in its own way so always follow the instructions carefully. Most of these stencil films are supplied only by the roll and they are quite expensive.

Poster using cut-film stencil (38 × 50 cm)

Making a reversal stencil

1. On the inside of the screen:
● paint the image in oil-based ink and allow it to dry *or*
● draw heavily with a wax crayon *or*
● lay textured objects under the screen and make rubbings on the mesh with a wax crayon. (Be very careful or you will damage the mesh!)

2. Hold the screen with the top tilted away from you. Coat the screen with water-based filler by pouring a little into the taped area at the bottom of the screen and scraping it upwards with a piece of card or a small squeegee. Repeat until all the screen has been coated (⇨ (a) opposite).

3. Leave the screen to dry lying flat but propped away from any surfaces. (A fan-heater will speed this up.)

4. Check the screen for any pin-holes in the filler; if there are some give it a second coat of filler or spot them out with a brush and allow to dry.

5. Lay the screen down on a pad of newspaper and pour on a little screen wash. Rub the image areas with a cloth or a small brush until all the oil-based filler has been removed. (If it is stubborn, stand the screen vertically and with two rags rub both sides of the mesh at once, ⇨ (b) below).

6. When the open parts of the stencil are completely free of oil-based filler the stencil is ready to be used with oil-based inks.

(a) *(b)*

Using a water-adhering cut film

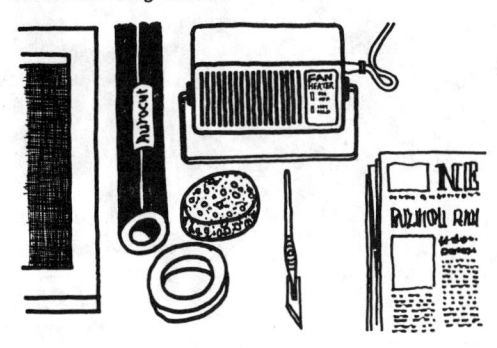

1. Draw out your design to the finished size on a sheet of paper. Tape it down on the table. Cut off a piece of film from the roll about 4 cm (1½ in.) larger all round than the area you are going to print. Lay the film matt side upwards on top of the design and hold it in place with tape at the corners.

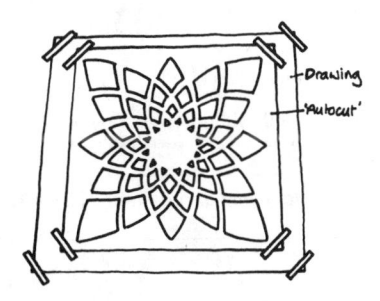

2. With a scalpel, cut round the shapes that are to print and remove the film from these areas by peeling it away. This cutting must be done very lightly or you will cut through the backing sheet, making it buckle, which may cause problems when you try to stick the stencil to the screen. (Practise first on a small piece of film before you start to cut out the stencil.) And remember again . . . the areas that you cut away are the parts that will print.

3. Untape the stencil and lay it matt side up on a raised pad or drawing-board. Position the screen on top of the stencil with the image in the centre of the screen.

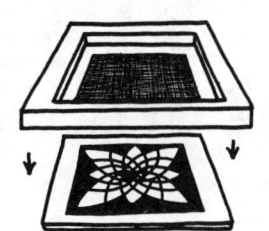

4. Get someone to hold down the edges of the screen so that there is close contact between the mesh and the stencil. Wipe the stencil and mesh gently with a moist sponge; the colour of the film darkens when it is wet and has stuck to the mesh. Be very careful not to get the stencil too wet as this will dissolve it too much and cause fuzzy edges.

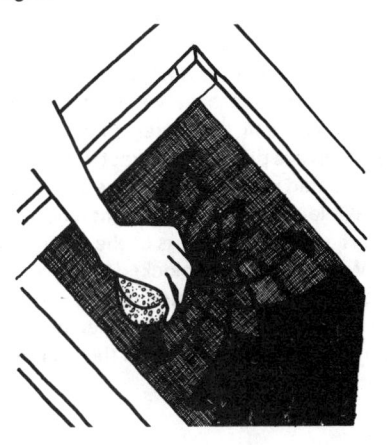

5. Only wet an area of about 15 cm (6 in.) square at a time and then blot immediately by laying newspaper on it and pressing gently with the flat of your hand.

6. Leave the stencil to dry for twenty minutes with a warm fan blowing on the inside of the screen (or overnight without the fan). When the stencil is dry, peel off the backing sheet. (Try a corner to see if it is dry; if the backing sheet does not come away easily or if it brings the stencil with it instead of leaving it stuck to the screen, it is not dry, so leave for a while longer.)

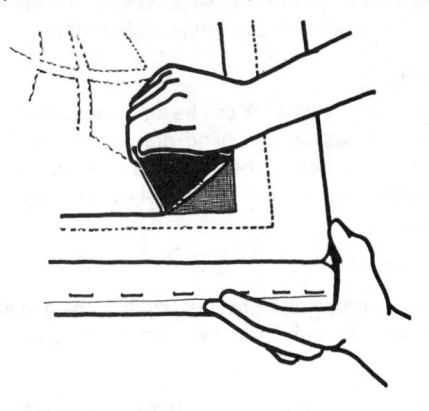

Using a lacquer-adhering cut film. This is basically the same process as for the water-adhering films, but the film is attached with the appropriate adhering liquid, and with some brands is then ironed on. Getting this type of stencil to adhere may sometimes be a tricky process: you should practise first with a small piece.

SCREEN PRINTING

MAKING PHOTO-STENCILS

If you want to print small words (that would be too fine to cut out as a cut-film stencil), photographic images or other more complex designs you may need to use a photo-stencil. It is more expensive to produce stencils photographically than in any other way. This is because the materials are costly, and because to produce them you may need to build some specialized and expensive equipment (if you cannot find some to use, ⇨ page 58).

In this method your design is drawn, painted or photographed on to a transparent (or semi-transparent) film. This is called the 'positive'. The positive is placed in contact with a 'presensitized film' or a screen coated with light-sensitive emulsion, and exposed to ultraviolet light. Where the light reaches the sensitive film or screen through the clear parts of the positive, it is hardened; and where the light is blocked because of the opaque parts of the positive the film or coated screen remains soluble. These blocked opaque parts are then washed out to make the open areas of the stencil, through which the ink passes to make the print.

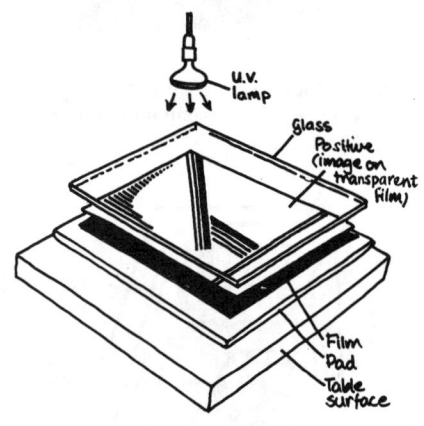

Presensitized film being exposed to ultraviolet light

In this section these processes will be described in this order:
1. The positive.
2. The light source.
3. Producing the photo-stencil.

If you are not familiar with photographic methods you may find all this initially a bit confusing. We suggest you read the whole section first to get a general idea before looking at the details.

THE POSITIVE
For simpler designs the artwork can be made directly on the transparent film. This is called a 'hand-drawn positive'. More-complex designs may need to be drawn on paper and then photographed on a process camera to make a 'photographic film positive' (⇨ page 102).

Hand-drawn positives

1. Draw out your design on a sheet of paper to the actual size of the finished print (or rough it out if you are going to use instant lettering).

2. Tape a piece of tracing film in position, matt side up, on top of your design. Tracing film is specially manufactured for this type of work, and comes by the roll.

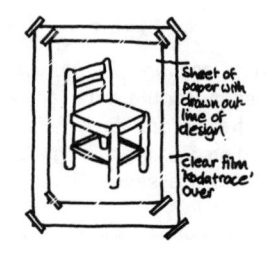

Two-colour festival poster using photo-stencil (51 × 37 cm)

One-colour poster by children on an adventure playground using hand-painted positive and photo-stencil (75 × 50 cm)

MAKING PHOTO-STENCILS 47

One-colour festival poster using hand-painted positive and photo-stencil (30 × 59 cm)

3. Trace your design on to the film in one of the following ways:
- 'Process black', 'photographic opaque' or other black ink or paint can be used with either brush or pen.
- Instant lettering can be rubbed down on to the film in the usual way for dry-transfer lettering (⇨ page 94), but do this on a hard surface or the letters will crack. Draw your guide-lines on a sheet of paper placed under the tracing film. The lettering can be made on separate pieces of tracing film and then Sellotaped into position on the main sheet, if that is easier.
- A grease pencil, like a 'Chinagraph' crayon, can be used to draw on the film if it is applied thickly so that it is sufficiently dense.
- Black-paper shapes can be taped on to the tracing film if large areas of flat colour are needed.
- 'Masking film' (such as Amberlith or Rubylith) is specially made for cutting out positives. It is very similar to the hand-cut stencil films mentioned earlier (⇨ page 44), being made up of a thin layer of film on a transparent plastic backing. The film is of an amber or ruby colour, which does not let the ultraviolet light through. Use it by laying a piece of the film over your design and cutting around the shapes that are to print. With this type of film the parts of the film which you *do not peel away* are the areas that will pick up on the stencil. Masking film is useful where you have a design with many straight edges, because it is easier to cut straight lines than paint them!

These methods of making the positive can be combined. A clear adhesive tape like Sellotape or Scotch tape can be used to tape pieces of tracing film together if some parts of the design have been made in one way and some in another.

4. When you have finished making the positive, check it over a light to make sure that all the areas are opaque to light. If necessary touch up with a little opaque ink.

Photographic positives. If you have access to a dark-room and have a little experience in photography it is possible to turn your own photographs into film positives suitable for screen printing. The normal negative can be put in an enlarger and exposed on to special film to break it up into tones and patterns made up only of

One-colour concert poster photographed from artwork and using photo-stencil (50 × 36 cm)

48 SCREEN PRINTING

black shapes on clear film. Two types of film do this, line film and halftone film.

1. *Line film.* This film only picks up black or white, so all the middle tones are lost, but this creates a strong image which is very suitable, for example, for posters. You can produce this result yourself by exposing a normal negative on to a piece of line film in the dark-room, or it can be done on a process camera (⇨ Glossary), in which case make sure that you ask for a 'line positive'.

One-colour poster using photo-stencil showing use of a photo converted into line (36 × 50 cm)

One-colour poster using halftone photograph and photo-stencil (90 × 57 cm)

2. *Halftone film.* With this film the different tones of grey in the negative are converted into a pattern of tiny black dots that give the impression of different shades or tones. The best way to get a halftone positive made is to get your photograph 'screened' on a process camera (⇨ page 102). It can also be done, more cheaply, by exposing your negative through an enlarger on to Autoscreen film, which has a dot structure already built in (for details ⇨ page 99). You must decide what size dot you can use on the screen you have, defined by the number of dots to the inch. It is not advisable for a beginner to try to print anything with a screen finer than 45 dots to the inch: screens finer than this can be very difficult to print. The best size to start off with is about 20 to the inch. Your dot size should be at least three and a half times bigger than your mesh count (⇨ page 36); if the dots are too small there won't be any threads on the mesh for them to rest on! (So if you are using a 77 mesh your halftone should not be finer than 55 dots to the inch.)

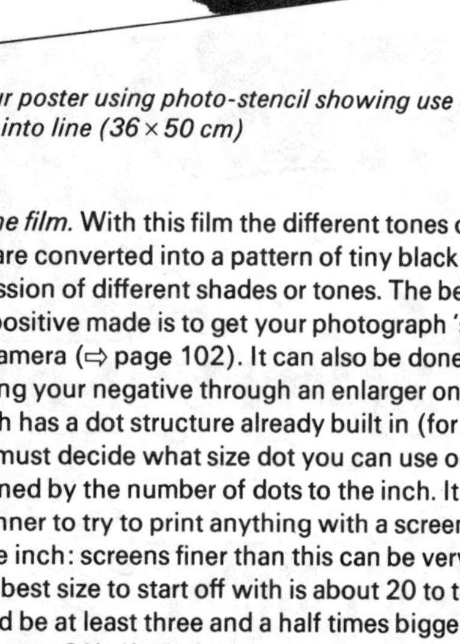

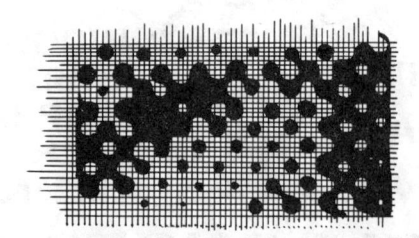

Close-up of dot size in relation to mesh threads

Making positives for multicolour work. If you are printing a design with more than one colour you will probably need to make a separate positive and stencil for each colour in the design. These stencils are usually put on to separate screens; they can, however, go side by side on one screen if they are small and your screen is large enough.

The main difficulty with multicolour prints is that of getting all the colours to fit together accurately when you print them. This is called 'registration'. There are things you must do when making your positives if you want to get accurate registration in the final print:

1. After you have made the positive for the first colour, leave it taped down in position on top of the design. Put another layer of tracing film on top of the first and tape it down, and on this make the positive for the second colour, tracing through from your design underneath. This can be repeated for as many colours as there are in the design. (Leaving the positives in position on top of each other means that you can see exactly how all the overlays fit together, and that they are accurate.)

2. Make 'registration marks' (small crosses) at the centre of the top and bottom of each positive as you make it. You later line these marks up with each other to ensure accurate registration when you are printing.

3. Wherever two colours are to meet up in the design, allow an overlap of about 3 mm ($\frac{1}{8}$ in.) of the one colour over the other. This gives a bit of leeway, so that if you cannot get the registration exactly right you will not have white lines between two colours that should meet up.

These instructions for making multicolour prints refer to making positives, but they also apply if you are *cutting out* stencils for

Photo-stencil for first colour

Photo-stencil for second colour

Three-colour poster using two photo-stencils and a hand-cut stencil for the background (37 × 51 cm)

several colours. (If you need to make your paste-up on paper for it to be photographed on a process camera read the section on multicolour paste-ups in the Design and Paste-up chapter, ⇨ page 85.)

THE LIGHT SOURCE

To make the stencil the positive is laid against the sensitized film (or coated screen) and the light is directed on to it. The lights must be rich in ultraviolet. Bulbs such as Photoflood No. 2 are the cheapest, but mercury-vapour lamps, which are stronger in ultraviolet and permit shorter exposure times, are more commonly used.

A simple light source can be made by suspending the bulb above a table-top. Check the manufacturer's instructions for the recommended distance between the bulb and the positive. (A distance of 50 cm (20 in.) is usually recommended for mercury vapour.) The sensitized film or coated screen is put under the light on the table-top to make the exposure (⇨ pages 97–8). At a height of 50 cm (20 in.) one mercury-vapour bulb will be enough to expose stencils up to a size of 50 cm (20 in.) square; if you are making larger stencils you will need to build a unit with two bulbs.

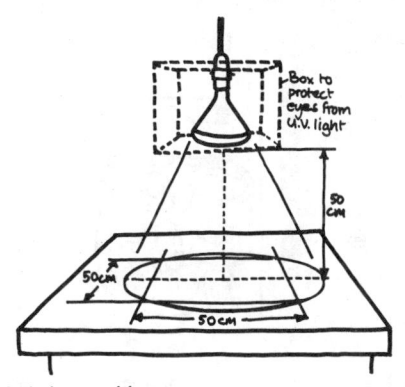

Bulb suspended above table-top

A light-box. If you have the space the lights can be fitted into a 'light-box'. The sensitized film (or coated screen) is placed on the glass top (of the box) to be exposed. (The light-box can be built with an ordinary bulb in it too, so that you can also check positives on it.) It can be made out of an old tea-chest or packing crate, but the diagram shows a more elaborate method.

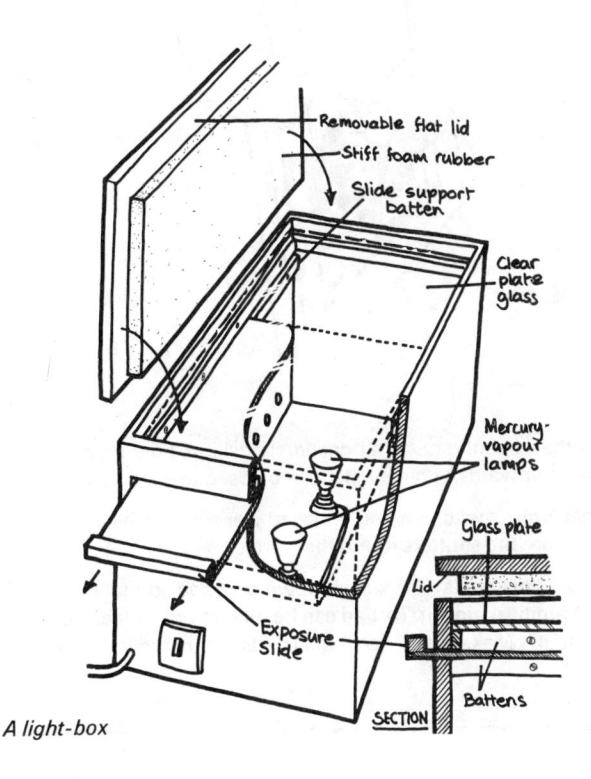

A light-box

Industrially made light-boxes incorporate a vacuum in the lid to ensure very tight contact between the positive and the stencil, but this is only absolutely necessary for making stencils with very fine halftones. For most types of work a sheet of glass will hold the positive tightly enough.

Whichever way you make your exposure unit it should incorporate a number of features, It should:

- Keep the bulbs at a fixed distance from the surface on which the sensitized film or screen will be laid.
- Enable (a) the light to be covered while the stencil is positioned and (b) the light to be uncovered for the exposure to be made once the bulbs have warmed up.
- Hold the positive firmly up against the sensitized film or coated screen during the exposure.

PRODUCING THE PHOTO-STENCIL

The photo-stencil may now be made. This is done by exposing the positive either directly on to a screen coated with a light-sensitive emulsion or on to a presensitized film which is later stuck to the screen. You must decide which method suits you best before you build your light source, since it will affect the way you build it. (The instructions that follow assume that you are using a light source from above; if you are using a light-box you will need to reverse the relative positions of positive to screen or film).

Making and using coated screens (such as Seriset)

1. Degrease the screen and allow it to dry (⇨ page 40).

2. In dim light coat the screen with emulsion. Pour a little into the taped-up areas at one end of the screen with the screen held upright, and pull it up the screen with a squeegee in one smooth movement. (Look at the manufacturers' instructions to see whether one coat is required on each side of the mesh.)

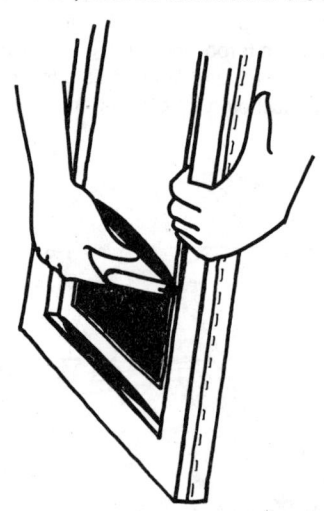

3. Lay the screen to dry in a horizontal position, printing side facing downwards. A fan-heater can be used to speed up drying.

4. Meanwhile, switch on your bulbs to warm up for five minutes, making sure the shutters round them are closed.

5. Still in dim light lay the screen down on a pad so that the mesh is lying firmly against it. (A pad can be made from a block covered in thin foam. Make it of a size that just fits within the inside of the screen.)

6. Lay the positive face down on the mesh (so that it reads back to front) and Sellotape it in position. Place a sheet of glass over all this to hold the positive closely against the mesh. Make sure this is all centred under the light.

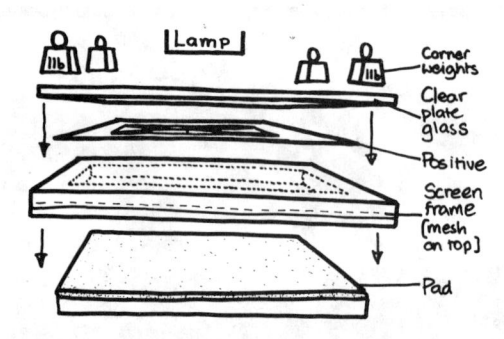

7. Make your exposure by opening the shutters on the lights. (When you first make a coated screen do a test exposure on to the screen to check that the manufacturer's recommended time is correct for your light system.)

8. When the correct exposure time has passed switch off the lights, lift off the glass, and untape the positive from the screen. Put the screen in the sink and spray gently with cold water. After one or two minutes increase the pressure and continue spraying until all the open parts of the stencil are clear and the water runs off clear.

9. Leave the screen to dry in a horizontal position for a couple of hours or fan-dry with warm air. When dry check for pin-holes (tiny spots revealing a hole in the emulsion) and use a little water-based filler to spot them out.

MAKING PHOTO-STENCILS 51

1. Coated screens. In this method the screen is coated with a light-sensitive emulsion. There are many types of these available from specialist suppliers, so follow the specific instructions that relate to the product you buy. Coated screens are much cheaper than the presensitized films described later (about one-third of the price), but they are not always practical, because the screens, which are affected by light, have to be coated in a dimly lit room (e.g. with only a 40-watt bulb or behind drawn curtains) and if coated in advance have to be stored in the dark.

These stencils are very strong (thousands of copies can be printed from them) and may be used with either water- or oil-based inks.

2. Presensitized films. There are many brands of presensitized films available but they all have to be bought by the roll and all are expensive. The various brands have their own methods of developing, using a variety of chemicals, and need washing out at different water temperatures. Some brands are simpler to use and only need to be washed out in cold water. These brands are among the most expensive but they can be very useful to the beginner working without sophisticated equipment.

Using a presensitized film (in this case, Autostar). Proceed as follows (manufacturers' instructions will vary):

1. Switch on the lights to warm up for five minutes, and degrease the screen (⇨ page 40); leave the screen wet.

2. In subdued light cut out a piece of stencil film 3 cm (1¼ in.) larger all round than your positive.

3. Lay the stencil film shiny side up on the foam-covered block. Lay the positive matt side down (reading back to front) on top of it.

4. Lay a sheet of glass on top of both so that the stencil and positive are kept in close contact. Make sure that this is all positioned centrally under the light.

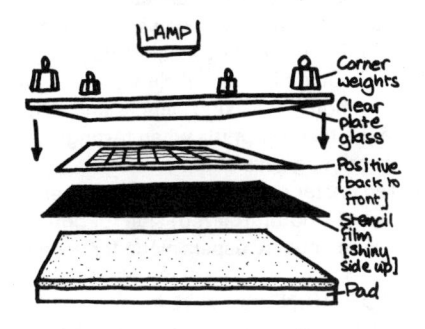

5. Make the exposure by opening the shutters. (The recommended exposure time for Autostar is two and a quarter minutes: do a test beforehand to check that this is correct on your equipment.)

6. Lift the glass and remove the positive and stencil film from under the glass. The stencil is now exposed. Pin the stencil film shiny side *down* on the upright board in the sink and spray it evenly all over with cold water until the colour of the film has washed out of the open areas and they are completely clear. Continue spraying for a couple of minutes more.

7. Dampen the screen with cold water (unless it is still wet from degreasing). Unpin the stencil and place it emulsion side down on to the under side of the screen. The best way to do this is to hold the screen at an angle, put the top edge of the stencil in position, and then slowly bring the screen to an upright position. The stencil will then fall into place with no air bubbles.

8. Lay the screen face down on a sheet of newspaper, and blot the inside of the mesh very gently with the flat of your hand or a small roller until all the excess moisture has been absorbed from the stencil.

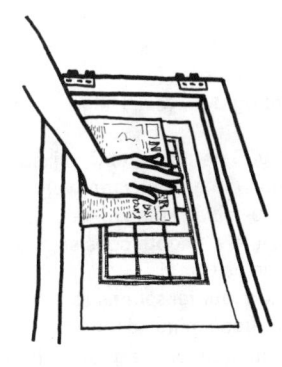

9. Allow to dry for ten minutes in front of a warm fan (or overnight). When the emulsion side of the stencil feels dry to the touch, peel off the backing sheet. (Try a corner; if the backing brings the stencil with it instead of leaving it stuck to the screen, then it is not yet dry.)

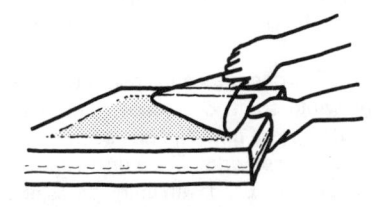

MAKING THE PRINTS

INKS

There is a great variety of inks, of which the ones most commonly used for simple printing are either water-based (mixed up and cleaned off with water) or oil-based (mixed up and cleaned up with white spirit or other solvent).

1. Water-based inks. These are found most frequently in the form of fabric-printing dyes, which can be used on paper as well as fabric. They are cheap and easy to use for beginners because they do not require expensive solvents and can be cleaned up with water. As they are transparent they will make 'overprints'; for example, if you print red on top of blue you will get purple. Fabric dyes can only be used with stencils that are *not* water-soluble (cut-paper, lacquer-adhering cut films, coated screens etc.), because the water in them would dissolve a stencil that washes off with water.

Bought fabric dyes. You can buy fabric dyes from various manufacturers. They usually come as a binder into which you mix the colours yourself to get the required colour dyes. You can also buy ready-mixed fabric dyes, but they are much more expensive.

If you are using a binder-and-colour system mix it up by adding a few drops of the colour to the binder in a jam jar until you get the required colour. A full jam jar is enough to print approximately thirty prints. Prints on fabric must be ironed for five minutes (with an iron set for cotton) once the print is dry; this is to fix the dye so that it will not wash out. Screens and equipment can be cleaned with cold water as soon as you have finished printing. Mixed-up dyes can be stored in airtight containers for several months. Note that some of these products do contain some white spirit or mineral spirits, as well as water, so be careful if you try to use such oil-based stencil types as wax crayon, which may be dissolved by the dye.

Fabric dyes certainly can be used to print on to paper (they give very colourful results), but the prints are not waterproof, and the paper may crinkle up a bit.

Home-made water-based inks. For making prints on paper you can manufacture your own inks at home by mixing colour into a binder (to make it the right consistency, like thin custard).

For *the colour* you can use:
- Special fabric-printing dyes (e.g. Helizarin colours or other bought fabric dye).
- Household dye (e.g. Dylon or Tintex), mixed up into a concentrated solution with a little warm water.
- Poster paint (the pre-mixed kind works best).
- Powder paint: mix it to a smooth consistency with a little water.

For *a binder* you can use:
- Wallpaper paste: add four teaspoons to half a litre (17½ fl. oz) of cold water, leave for fifteen minutes then stir till smooth.
- Specially made printing binder (e.g. Colorflex or Manutex): follow manufacturer's instructions on mixing and using.

To mix the inks, fill a jam jar with one of the binders from the list and mix in some colour until you get the strength and consistency you want. (With these transparent colours the best way to see exactly what the colour looks like is to dab a little on to the paper with your fingers.) Home-made inks will not keep for more than about a week, when the starch-based binders start to rot.

2. Oil-based inks. The cheapest and easiest to use of the oil-based inks are the white-spirit- or mineral-spirit-based types. They make an opaque waterproof print and can be used on most types of paper, card, hardboard or wood. They dry in about fifteen to twenty minutes and can be used with many types of stencil, but not with oil-based stencils (e.g. lacquer-adhering hand-cut, wax crayon, etc.), which would be dissolved by the ink. If you need a transparent colour this type of ink can be used with an 'extender base'. Oil-based inks are thinned down with white spirit or mineral spirit (or their appropriate solvent) until they are the consistency of thin cream, and they are cleaned up after printing with screen wash (a stronger solvent). Other types of oil-based inks are available if you need to print on surfaces such as glass or plastic; for more details see the manufacturer's technical information.

PAPER

You can print on almost any paper, but watch out for paper that is highly absorbent or very glossy, on which it might be difficult to print. In practice you will find that the type of paper you use depends more on what is available than anything else.

Buying paper. It is expensive to hold stocks of paper unless you are printing regularly; you could try:
- Buying in bulk from shops that will give a discount on larger amounts (more than 100 sheets).
- Buying 'left-overs' from a local printer (who may also be prepared to cut up paper for you into smaller sheets).
- Buying through your local-government supplies depot. You should be able to do this if you are a local-government organization, such as a school, youth club or play centre. This is the cheapest source, for those entitled to make use of it, although the range of paper stocked may not be very wide.

What kind of paper? The two most useful types of paper you need are:
- *Newsprint:* the cheapest paper obtainable. It is often possible to buy or obtain ends of the roll from a newspaper printer. Newsprint is a very thin paper, but this can be an advantage if you want to fly-post posters printed on it. You will find that you will need plenty of newsprint both for making stencils and as rough paper on which to take the first few prints when testing, to make sure that everything is printing satisfactorily.
- *Machine-glazed poster paper:* often referred to as MG. This is a white paper with one smooth (glazed) side for printing on. It is the cheapest of the better-quality papers made for poster printing.

PLANNING YOUR PRINTING AREA

One of the nice things about screen printing is that you can do it virtually anywhere, if you have to: the kitchen table, the floor, outdoors, etc. But if you think you will be printing regularly and you have a permanent space available, it is worth setting up the room so that you can use it in the most efficient and practical way – you will save time and energy!

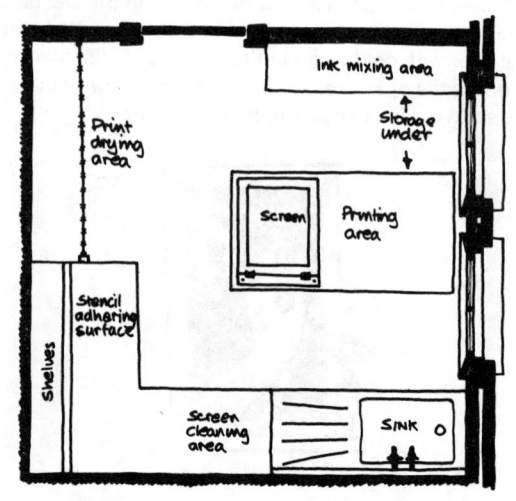

One way to arrange a screen-printing workshop

Here are some of the things you will find useful:
- Smooth, flat-topped table to use when attaching stencils.
- A surface to use when mixing up inks (the top of a cupboard?).
- Storage for inks, paper, prints and screens (in cupboards, plan chests, on racks or shelves?).
- A sturdy table to use the baseboard on.
- A surface on which to clean screens.
- A sink large enough to stand your screens up in, with a splashboard at the back, and hot and cold water with a shower attachment.

PREPARATION AND REGISTRATION

Before you start to print it is important to have everything ready and close at hand; once there is ink in the screen there may be problems if you have to stop to cut up your paper or mend a stencil, because the ink will start to dry and clog up the screen.

1. Check the stencil for any pin-holes or other gaps and touch up with filler (or with small pieces of Sellotape put on the underside of the mesh, if you are only making a few prints).

2. Check that your squeegee is at least 5 cm (2 in.) wider than the area that you are going to print.

3. Cut your paper to size and stack it beside the printing table.

4. Prepare your ink: water-based ink should already be of the right consistency (like thin custard), but oil-based inks will need thinning with white spirit until they are like thin cream.

5. Have the following within easy reach: a supply of rags or kitchen roll, newspapers, a bin for rubbish and dirty rags, Sellotape, turps and screen wash (or sponge and water if using water-based inks).

Registration. This is the name given to the process by which you ensure that all the prints are made in exactly the same position on each sheet of paper. This is very important if second or subsequent colours are to match up to your first colour accurately. Pieces of tape or thin card (called the 'stops') are used to make sure that the paper is placed in exactly the same position for each printing.

1. Find the correct position on the baseboard for the sheet of paper, by bringing the screen down and looking through the stencil. If it is the first colour (or a one-colour poster) position the paper so that the final print will be properly centred on the paper. If it is the second colour then move the print around until the second colour is lined up with the first colour and the registration marks match up.

2. Once the paper is in the correct position stick the stops down on to the baseboard so that they touch the two edges of the paper as shown.

3. If you are printing more than one colour, make a note of which edges were used for the registration (by drawing crosses in pencil to show where the stops were placed) and always use the same two edges for your registration or it will not be accurate.

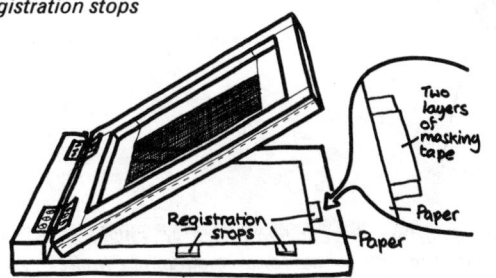

Position of registration stops

Taking a print

1. Place a sheet of print paper in position against the 'stops' on the baseboard, and lower the screen.

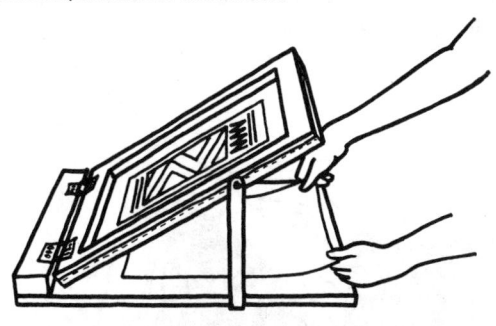

2. Pour ink into the taped area at the far end of the screen. Take the squeegee in both hands and collect up the ink in front of it.

3. Make one firm even pull towards you with the squeegee tilted at an angle of about sixty degrees.

4. When you reach the other end, lift the screen a couple of inches with one hand while holding the squeegee where it is in the other hand (to prevent the ink from running back down the screen) and allow the print to fall away or peel it away if it does not fall.

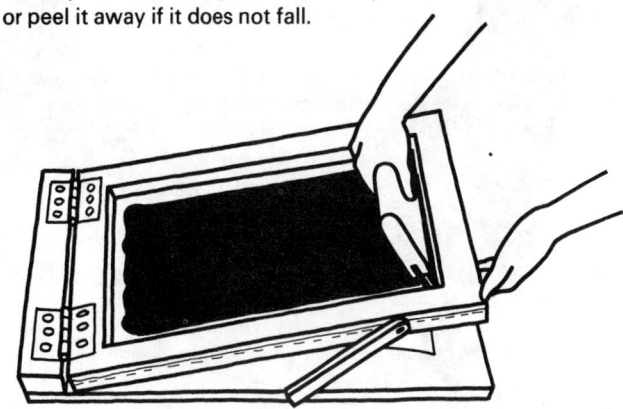

(continued overleaf)

(continued from previous page)

5. Then push the ink back down the screen while it is still raised up, holding the squeegee at the same angle as the forward pull but this time without using any pressure. This is called 'flooding back'. It helps to prevent the screen from drying up and blocking in so far as a small amount of ink is left in the open parts of the stencil until the next print is made.

6. Put the print in the rack to dry. Position the next sheet of paper and continue to print your sheets.

PRINTING MORE THAN ONE COLOUR

If you want more than one colour on your print the usual procedure is to make a stencil for each colour in the design, print the first colour, allow it to dry, print the second colour, allow it to dry, and so on. Multicolour prints are expensive because they use more ink, more stencils and more time! One way to save money and time is to find a way to print all the colours in one go. There are two ways to do this:

1. *'Rainbow' printing.* By pulling a few colours at once across the screen with the squeegee you can get a very nice 'rainbow' effect. Pour small amounts of the colours you want into the top of the screen and pull the squeegee across the screen in the usual way. The first prints will be a bit stripey but after several prints have been made the colours start to blend together, producing a more subtle effect. This method works best on a run of about thirty to fifty prints, beyond which the colours tend to go muddy even when you add more colour into the screen. This effect looks most dramatic when you are printing large areas (like sky backgrounds) but it is equally useful as a way of printing words in lines of different colours on a poster.

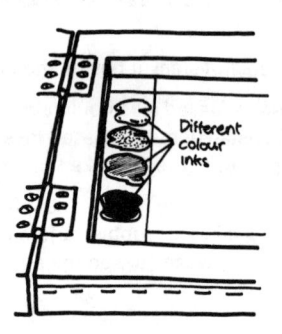

2. It is also possible to build card 'walls' down the length of the screen to keep different colours of ink separate. Using a different squeegee for each colour, you can print first one colour then the next, and then change the paper ready for the next print. This method is only suitable if you design your print with easily divided areas of colour.

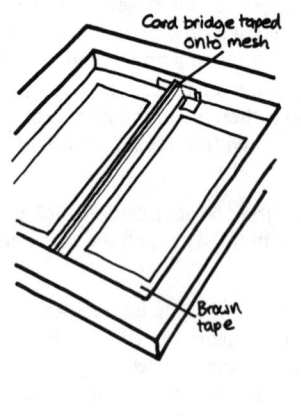

Festival poster using cut-film stencil – the blank space has been left for writing in specific events (one-colour rainbow effect, 75 × 50 cm)

CLEANING UP

When you have finished printing remove all the ink from the screen and the squeegee as soon as possible – if it dries it will be much more difficult or even impossible to get out of the mesh.

1. Lay some newspapers under the screen, scoop up as much ink from the screen as you can with a piece of stiff card and return it to the jar. (Mixed-up oil-based inks keep well in screw-topped jars.)

2. Wipe up the rest of the loose ink with a rag (or with a sponge and water if using water-based ink).

3. If the stencil is loose (paper, for example) remove it now.

4. *Oil-based ink:* Lay the screen down on some newspaper and pour on screen wash. With a rag loosen the ink in the screen; change to a clean rag (or sheet of newspaper) and soak up the dissolved ink. Repeat the process until all the ink has been removed from the screen.

Water-based ink: Hose the screen with cold water until all the ink has been removed, or lay the screen down on newspapers and clean with a sponge and water.

5. Any stubborn ink should be removed by rubbing the mesh from both sides at once with two rags soaked in screen wash (or water if using water-based inks).

6. Dry the mesh with a clean rag.

7. Clean the squeegee and mixing knives with a rag and screen wash (or water with water-based ink).

8. Remove the stencil from the screen if you have finished using it. (Otherwise it can be stored until you need to use it again.)

OTHER USEFUL INFORMATION

PRINTING T-SHIRTS

It can be very easy to print T-shirts with portable equipment. Kids enjoy doing this as a project and it is also very effective for publicity or fund-raising. The process is basically the same as for printing posters but you must use dyes instead of inks. (You can in an emergency use poster inks but they will eventually wear off from the T-shirt.)

You will need:
- Screens: stretched with either cheap organdie or coarse nylon (no finer than 43 threads to the centimetre).
- Squeegee: rounded blades rather than square ones are best for printing on fabrics, which are more absorbent than paper as they allow more ink to go through the screen, but any type will do.
- Fabric dyes: either the binder type or ready-mixed ones.
- Stencil: you must use a type that will not be affected by the water-based dyes. Any of the following are suitable:

Cut paper (⇨ page 41).
Liquid filler, such as oil- or cellulose-based ink (⇨ page 43).
Hand-cut film, such as lacquer-adhering types.
Photo-stencils using a coated screen (⇨ page 46).

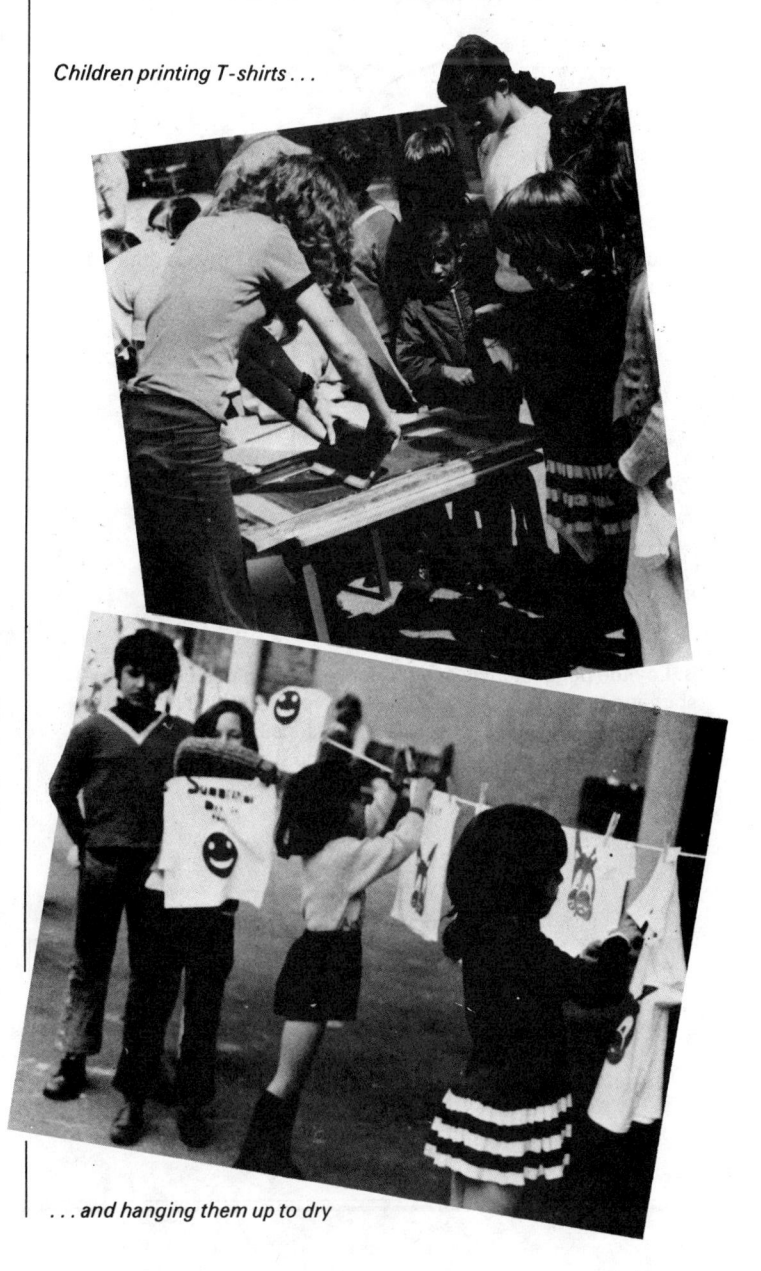

Children printing T-shirts . . .

. . . and hanging them up to dry

56 SCREEN PRINTING

Printing the T-shirts with a paper stencil

1. Cut out a cardboard template to put inside the T-shirt to prevent ink coming through on to the back.

2. Put the template in the first T-shirt and lay it down on the table or baseboard.

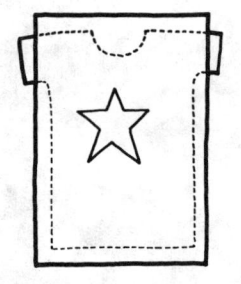

3. Lay the paper stencil, and any loose parts, in position on the T-shirt and bring the frame down on top very carefully. (Make sure that all of the T-shirt is covered by the stencil; if it is not you will have to lay more strips of paper under the screen to fill up the spaces.)

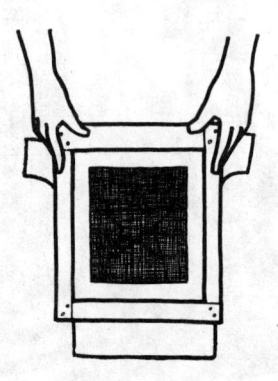

4. Get someone to hold the screen so that it does not move as you are printing (if it does, the T-shirt will smudge).

5. Pour some dye into the taped-up area at the far end of the screen, and make a firm pull towards you with the squeegee.

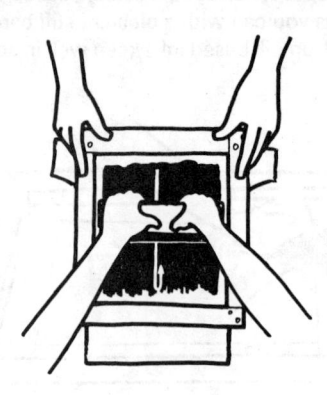

6. Make a second pull with the squeegee by pushing it back up the screen. (Some people find this difficult, in which case pick up the dye on the squeegee and carry it back to the far end and make a second pull towards you.)

7. Lift the screen very carefully, checking that the stencil has adhered to the underside of the screen. Tape the stencil to the corners of the underside of the screen to prevent it from falling away.

8. Get someone else (with clean hands!) to remove the T-shirt, take out the template and position the next T-shirt ready to print. (Look inside this first T-shirt to see how much dye came through on to the template... you may need to use an extra pull of the squeegee or you may find that lots of ink is coming through, in which case one pull would be enough.) All fabrics absorb differently so you should always check the first print to see how many pulls of the squeegee you need.

9. Hang the T-shirt up to dry and continue printing the number you need.

10. When the prints are dry they should be ironed for five minutes with a hot iron to fix the dye so that it will not wash out.

11. When you have finished printing clean the dye out of the screen immediately with *cold* water: if you do not do this it may be very difficult to remove (⇨ page 55).

OTHER USEFUL INFORMATION 57

Printing T-shirts with other types of stencils. An alternative method is to make your stencil, put it on the screen (as described previously), and then follow the instructions given opposite for printing T-shirts, ignoring the references to paper stencils since your stencil will already be attached. If you need to print a number of T-shirts and cannot do them all at the one time, it is better not to use a paper stencil, since this has to be removed when you clean up. If you use one of the other types mentioned you can first print a few T-shirts on one day, then clean up and print a few more on another.

Right: T-shirt to be worn by festival organizers, printed with cut-paper stencil

Far right: T-shirt printed to raise funds for a childbirth group, using cut-film stencil

Right: T-shirt to publicize a neighbourhood festival, using a photo-stencil

LARGER EQUIPMENT

Vacuum table. The vacuum table is the commercially built printing table used in the trade. Such tables are expensive to buy even second-hand, and people often wonder what advantages they have over simpler equipment. The following are the basic refinements of a vacuum table, which enables you to print long runs and more complex designs:

● The screen is clamped in position and counter-weights or springs hold it in the raised position while the paper is changed; this means that the print can be removed and the next sheet positioned quickly, saving time and energy if several hundred copies are to be printed.

● During printing the paper is held in position by the suction of the vacuum. This permits accurate registration, since as a consequence the paper will not move during printing.

● There is a space (which can be adjusted) between the mesh and the printing paper that produces what is called the 'snap-off', enabling you to make very crisp prints, essential when printing details such as those of fine halftone work. The mesh only touches the paper as the squeegee passes over it; it then springs away and this prevents the ink from spreading and blurring the edge of the image.

● Large areas (75 × 50 cm (30 × 20 in.) or more) can be printed with much less effort, particularly if you have a 'one-arm squeegee' unit. The squeegee is clamped to a universal joint which is attached to the arm that runs along the back of the table. A handle extends outwards at the front and the prints are made by pressing down on this arm and pulling it (and the squeegee) across the screen from one side to the other.

Disadvantages

● Vacuum tables are very expensive to buy new. They can be bought for about half price second-hand but will still cost several hundred pounds. Look in such magazines as *Exchange & Mart* or *Point of Sale* for ads, or talk to your local rep from the suppliers, who will probably be an amazing source of information about who is selling off equipment.

● The table takes up a lot of space which cannot be used for much else when it is not in use.

● You can build the table yourself, but this requires skill and would still be quite expensive (⇨ *Setting Up in Screen Printing* by Stephen Russ).

Drying-racks. If you are frequently printing long runs a professional, metal drying-rack can save time, space and energy: 200 prints can be racked to dry in an area 150 × 50 cm (60 × 20 in.) – you need only walk one or two paces, rather than back and forth to your washing line and pegs! Unfortunately, these racks are very expensive new; they also seem to keep their value, so that even second-hand they cost between one-half and three-quarters of their original price.

A metal drying-rack

Larger equipment can be expensive: whether it is worth getting will depend on how many prints you make, their size, how often you print and how complex the designs are. One way round the problem, if you have need of larger equipment, is to try to get permission to use other people's equipment. Try local colleges, adult institutes, community resource centres, print-making workshops, schools, and so on. You may also be able to buy various materials there. Alternatively, you might consider combining with other groups to buy the equipment collectively.

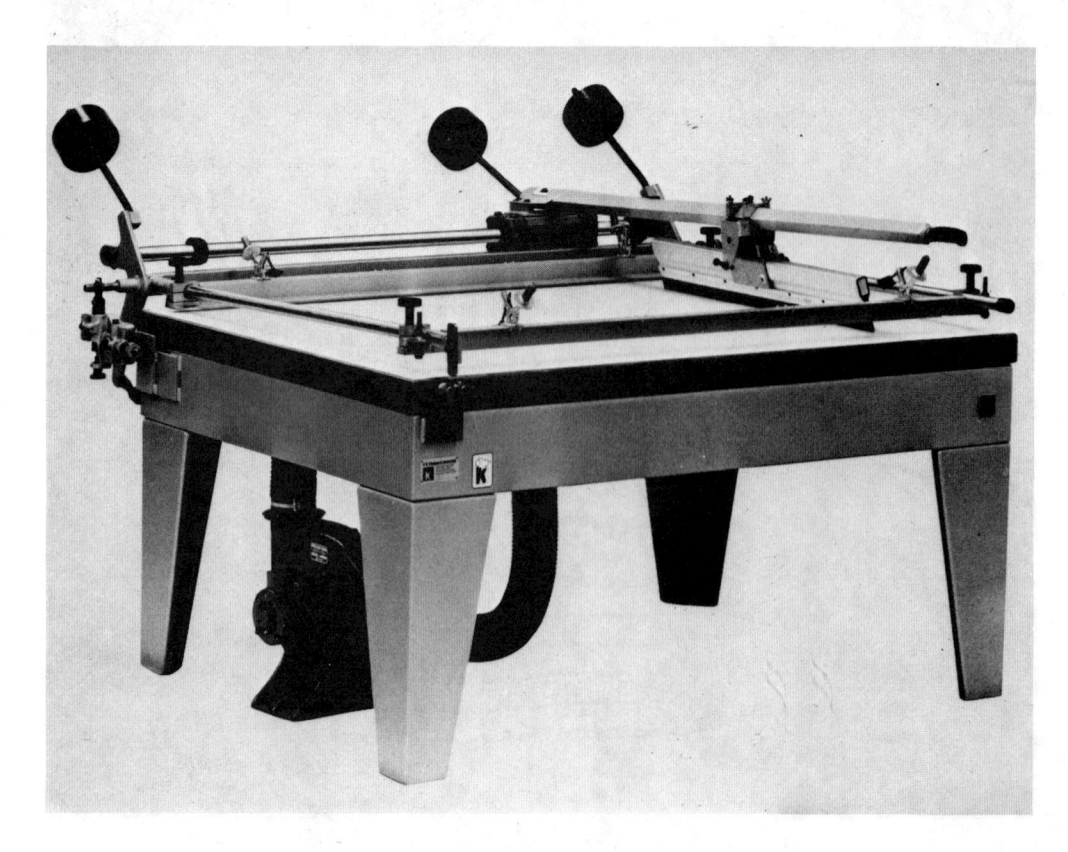

A commercial vacuum printing table (photo courtesy of Pronk, Davis & Rusby Ltd)

FINDING AND USING A COMMERCIAL PRINTER

To find a printer ask other people where they went or look in the yellow pages under 'Screen Process Printers'. Get a quote from two or three on the phone; tell them the length of the run, size, number of colours, percentage of ink coverage and whether the artwork is in line or halftone. Most printers will give a price, subject to seeing the artwork, so once you have done the artwork go to see them and get the quote verified.

When making artwork for a commercial screen printer it is best to supply black-and-white artwork done on paper with overlays for each colour (⇨ page 105 on paste-ups for more than one colour) and the printer will make the positives from this. You can save money by supplying your own positives (either hand-made or photographic), but make sure they are suitable and of a high enough quality: if not, the printer will rephotograph them anyway and charge you for it! You will probably have to wait two or three weeks for the printing to be done, but tell the printer when you hand in the artwork by what date you want it to be ready.

SPECIALIST SUPPLIERS

It is always useful to send off for the free product information that companies send out. You may be blinded by much of the technical jargon, but when problems arise it may be helpful. In any case you will need to buy a few products from some companies. It is worth getting in touch with the area reps – you may strike lucky and find someone who is interested in small non-commercial projects. Reps can be invaluable for technical advice, especially if you are just setting up a workshop; and they may often let you have plenty of free samples to try out..

Large specialist suppliers
The following sell everything for the screen printer:

U.K. SUPPLIERS

Dane & Co. Ltd
1–2 Sugar House Lane
London E15 2QN
Tel. (01) 534 2213

Graphic Display Products
Frogmore Road
Hemel Hempstead
Herts HP3 9RT
Tel. (0442) 63846

George Hallsales Ltd
Hardman Street
Chestergate, Stockport
Cheshire SK3 8DQ
Tel. (061) 480 5762

E. T. Marler
Deer Park Road
London SW19 3UE
Tel. (01) 540 8531

A. J. Polak
439–443 North Circular Road
London NW10 0HR
Tel. (01) 450 7747

Pronk, Davis & Rusby
90–96 Brewery Road
London N7 9PD
Tel. (01) 607 4273

Sericol
24 Parsons Green Lane
London SW6 4HS
Tel. (01) 736 8181

U.S. SUPPLIERS

Pearl Paints Co. Inc.
308 Canal Street
New York
NY 10013
Tel. (212) 431-7932

Naz Dar Screen Process Products
1087 N. Branch Street
Chicago, Illinois
IL 60622

Ulano Inc.
210 East 86th Street
New York
NY 10028

Screen Process Supplies Mfg Co.
1199 East 12th Street
Oakland, California
CA 94606

Suppliers of small quantities
The following will take orders for small quantities and supply kits for schools:

Selectasine
22 Bulstrode Street
London W1M 5FR
Tel. (01) 935 0768

Serigraphics
Fairfield Avenue
Maesteg, Glamorgan, Wales
Tel. (0656) 3171

Dyestuffs suppliers

Brico Commercial Chemical Co.
55–57 Glengall Road
London SE15 6NQ
Tel. (01) 639 2020
(Helizarin dyes)

Candle Makers Supplies
4 Beaconsfield Terrace Road
London W14 0PP
Tel. (01) 602 4031
(Prodion dyes, Manutex)

M. E. McCreary & Co.
815 Lisburn Road
Belfast BT9 7GX
N. Ireland
(Polyprint dyes)

COMMUNITY ACTION

FIGHTING BACK — ORGANISING IN THE 1980's

SPECIAL 50th ISSUE

FOCUS ON
housing campaigns ~
anti-nuclear campaigns ~
organising the unemployed ~ childcare projects ~

PLUS tenants news ~ adventure play ~ community arts

Cut the queues ~ ORGANISE NOW!

ALWAYS PACKED WITH NEWS AND INFORMATION ON

HOUSING · HEALTH · CUTS · CLOSURES
and much more

PRACTICAL ACTION NOTES, SUCCESSFUL TACTICS

The magazine for working class tenants and action groups, community projects, trade unionists & activists involved in the struggle

More information, free leaflets, subscriptions (£2.40) from PO Box 665, London SW1X 8DZ

Price 40p   NOW AVAILABLE

Leaflet to advertise a magazine (A4)

Offset-litho

Introduction 61

When to use offset-litho 62

Choosing a printer 62
Community printers 63
In-plant printing 63
Instant printers 63
Larger commercial printers 64

Planning 64
Size of paper and image 64
Size of press and quality of image 64
Length of run 64
Plates and quality of image 65
Paper 66
Photographs 66
Colour printing 66
Finishing 68

Keeping a small job simple, quick and cheap 69

Check-list 69

Dealing with the printer 69

INTRODUCTION

The aim of this chapter is to explain in very simple terms how offset-litho works, to give guidance in choosing the best printer for your job, and to run through the points you need to bear in mind when planning and designing a job to be printed by this method.

Lithography works on the simple principle that grease and water do not mix. An image is made on a printing plate. The image retains greasy printing ink. The rest of the plate, however, is damped by water which repels ink. In this way only the inked image prints. Offset-litho (or 'offset' for short) is so called because the image is first transferred as a mirror image on to a rubber cylinder (called the 'blanket') and then printed the right way round from this cylinder on to the paper.

All offset-litho presses have three cylinders: the plate cylinder; the blanket cylinder; and the impression cylinder.

The plate cylinder. The printing plate (made of paper, plastic or a thin sheet of metal) is clamped around the plate cylinder. As the cylinders turn, first water (or 'fount solution') and then ink are applied to the plate by a series of small rollers. The ink sticks to the image but not to the rest of the plate.

The blanket cylinder. As the plate turns on the plate cylinder it comes into contact with the rubber blanket cylinder. The blanket picks up the ink image from the plate.

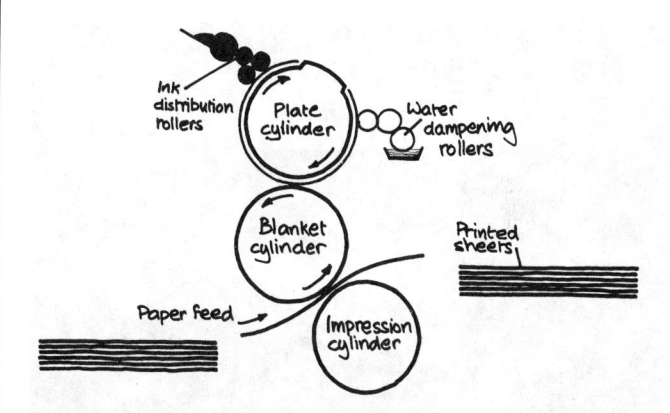

The three cylinders of the offset-litho press

The impression cylinder. Each sheet of paper passes between the blanket cylinder and the impression cylinder. The ink image on the blanket is transferred on to the paper as it is pressed against the blanket by the impression cylinder.

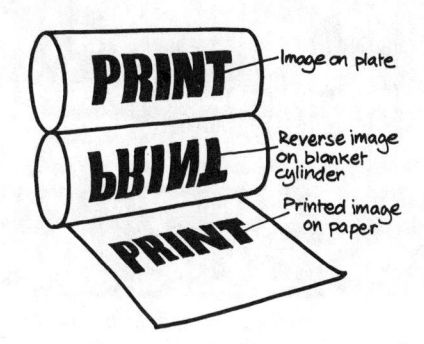

How the image is 'offset' on to the paper

An A4 leaflet to advertise a concert – making the paste-up yourself means that you control exactly how it looks

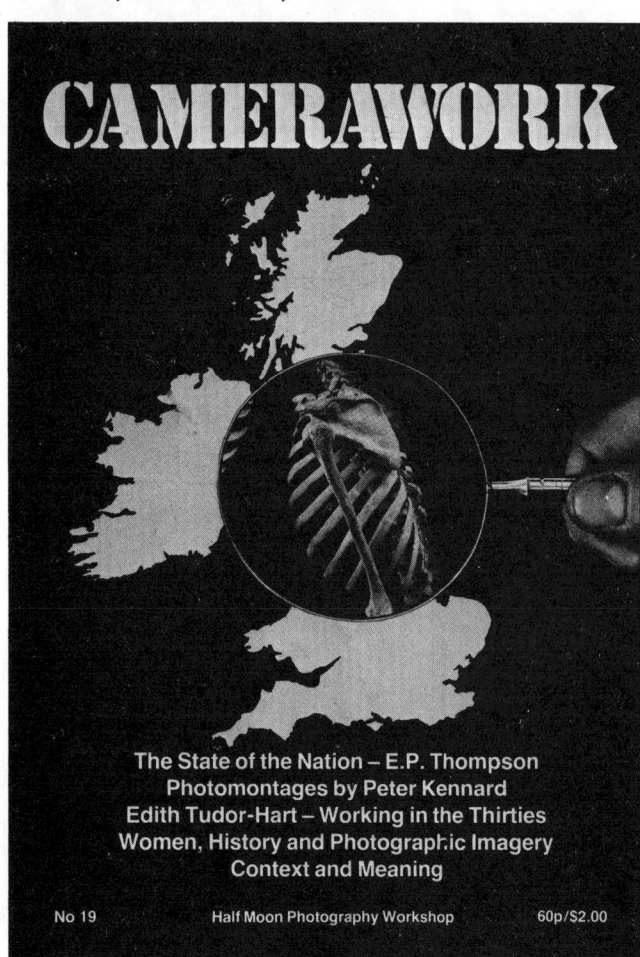

This magazine cover shows the good reproduction of photographs and high quality possible with offset

Offset-litho gives far better results than other cheap methods of printing, although the quality varies according to the equipment used and the skill of the operator. The customer also has a great deal of control over the end-product because he or she can produce the original design material (the paste-up) which the printer then photographically copies on to the printing plate. Photographs can be printed almost as easily as text (though they must first be converted into line or 'half-tone' dots, ⇨ page 97).

In the last decade or so the development of small offset presses that print on to A4 and A3 paper, and of 'instant' plate-making methods, have greatly changed the print industry. Offset-litho has become more accessible, flexible and cheaper, and has now replaced letterpress as the major method of printing.

These small offset presses have been developed mainly for use in offices. They are very easy to use (for this reason they are often called 'offset duplicators' by the manufacturers) and can do almost all that a larger press can do except print such good solid areas. Quick, simple plate-making systems for making plastic and paper plates have also been developed.

The development of all this technology and the variety of ways in which it has been taken up by commercial and non-commercial organizations has meant that print has become cheaper and more readily available to everyone.

WHEN TO USE OFFSET-LITHO

Of all the methods described in this book, offset is the best for quality, length of run, large paper sizes and accuracy of multicolour work.

Quality. Offset gives a high-quality print: crisp images, fine details, and good reproduction of solids and photographs, especially if a larger press (A3 or A2) is used.

Length of run. With runs of 500 or more, offset is often the cheapest method. Moreover, the longer the run the less each print costs.

Larger paper sizes. For anything larger than A4 you are likely to need offset, although you will find the larger presses only at big printers. Here is a rough guide (the metric dimensions of the various A sizes, and their equivalents in inches, are given in the Glossary):

Instant printers	A4, perhaps A3
Community printers, in-plant printing	A4, A3, perhaps A2
Large commercial printers	A2, perhaps A1
Web-offset printers	A2, A1 or bigger

Colour. Printing by offset enables you to use other colours as well as, or instead of, black, although there will be an extra charge for cleaning down the machine. Multicolour work can be printed to a high quality inasmuch as most litho presses can register colours accurately.

Cost. Compared with other methods, where you may only have to pay for the materials, offset may seem an expensive process. But it may be a necessary expense if you want a particular quality, colour or size, and it may, of course, save you time too.

CHOOSING A PRINTER

Most printers specialize in some way. It is important to find a printer who is used to handling your kind of job. Have a look at publications similar to your own, or ask other groups which printers they have used and would recommend. Look in the yellow pages (under 'Printers and Lithographers', 'Lithographic Printers' or 'Printers and Stationers'), resource directories and the printing-trade press, or ask at your local resource centre; there will be dozens of printers to choose from.

COMMUNITY PRINTERS

During the last few years quite a number of non-commercial print-shops have been set up. They may have different titles, calling themselves 'community', 'alternative' or 'cooperative' print-shops – or 'resource centres' – but they all aim to meet the print needs of their particular community. These printers usually have explicit aims, politics and publishing policies – for instance, they will often refuse to print anything they think is racist or sexist.

More traditional 'community' printers are the small presses, private presses or jobbing print-shops often run by individuals as a hobby or as a small business. The 'community' they serve may sometimes be more personal than that of the community print-shops, and they have been very useful in supporting small-scale publishing in such fields as poetry.

Community printers usually have small offset presses printing A4 and A3 work, and few have sophisticated plate-making equipment. Because their equipment is often second-hand the printing is sometimes not perfect, but these printers are cheaper than commercial presses and are experienced at helping to produce small-scale local publications that have a low budget.

Do not expect community printers always to work like a commercial printer. They may not be as fast or quite as efficient, say, as an instant printer, although if you have a rush job they will usually try to meet your deadline. However, they are often more helpful, showing you how to prepare your paste-up and sometimes how to do some of the printing yourself. It is important here to talk to the people who work in the print-shop and to do as much as you can to help.

Resource centres (usually funded by local authorities) sometimes have small offset presses. They tend only to provide facilities for groups in their own particular area, having a slightly different orientation from that of a community printer.

If there is no community print-shop (or resource centre) in your area it may be worth considering trying to set one up. Once a press of this kind is established many people discover the possibilities of print and are encouraged to produce their own publications.

IN-PLANT PRINTING

'In-plant printing' is the name given to printing facilities set up within an office or company. It may mean a small print-room or a very large printing department. Some in-plant print-shops will do work for outsiders but, since most concentrate on the needs of their own organization, personal contacts are often most useful here.

In-plant printing has developed fast over the last few years as small offset machines ('offset duplicators') and simple plate-making systems have been developed. This kind of equipment, which is designed to be used by office workers, is almost as cheap as stencil-duplicating machines, produces much better-quality prints and can print runs up to 10,000.

Many non-commercial organizations, such as schools, colleges, community centres and student unions, have also set up in-plant printing. Some of these are likely to make their equipment available to the local community both formally and informally. It is worth finding out what equipment your local institutions possess, to see if it is used all the time (many are very under-used) and then putting your case for community use to the management. For example, the town council might open up its printing department to the groups to which it gives grants, or the local school might let a community group print a newsletter on its equipment.

INSTANT PRINTERS

As small offset presses and instant plate-making systems have developed, so 'instant printers' have flourished on the High Street or on Main Street. They concentrate on producing cheap quick copies for offices and businesses. By standardizing machines, plate-making and paper size to A4 (and sometimes A3) they can produce short-run, black-and-white, single-sided work very quickly and cheaply. Most instant printers produce a detailed price-list so that you can see exactly how much a particular job would cost.

Front and back cover of leaflet produced by a community printer to advertise facilities and prices (A3)

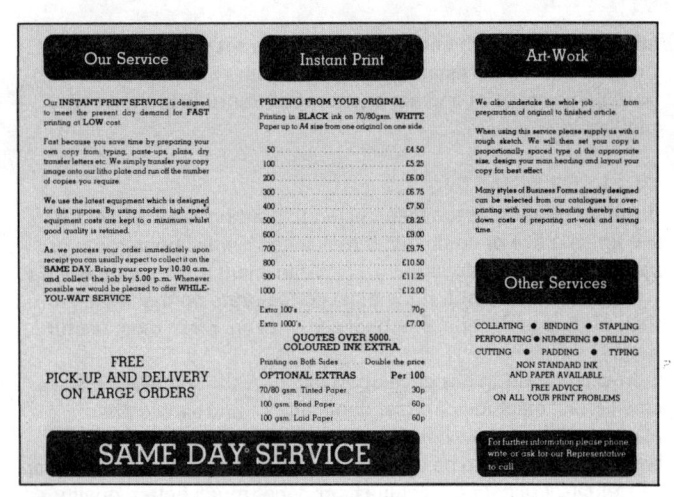

Instant printer's publicity leaflet (A4)

Except for short-run single-sided work, instant printers are usually more expensive than a community printer, but not necessarily much more so than other small commercial printers. Instant printers vary in the quality of their work – solids are often grey, and high-quality reproduction of photos is only possible if they use a metal plate (although reasonable quality can be obtained with a plastic plate if you use a screened photo-print with 65 or 85 dots to the inch – ⇨ page 98). Provide the printer with good clean artwork, for dirt and shadows may show up if paper or plastic plates are used.

Instant printers have made print much more widely available, but their prices are sometimes high and quality not always what it could be. It is worth using instant printers for the types of jobs they specialize in, but for other work or longer runs they are not the most economical printers to go to.

LARGER COMMERCIAL PRINTERS

Larger commercial printers have the large equipment, skilled workers and the commercial weight that other smaller printers do not have. They have a variety of sizes of press and can meet most print needs. Larger presses also enable them to print to a higher quality. They can often provide other services such as typesetting, design and finishing (⇨ page 68), and they also hold extensive ranges of paper. Large printers have to fit your job into a full schedule, with a corresponding delay in production, so obtain a firm delivery date from them and book your job well in advance.

In the U.K. large printers with unionized workers may refuse to print material that has been typeset or pasted up by non-union labour. This is part of the campaign by the unions to keep jobs in the face of technological change, and to stave off increased opposition by employers to the strength of the craft unions. Check on this point with the printer; it should not be too much of a problem, as most typesetters in the U.K. belong to the union, and if a paste-up has been done with 'voluntary' labour unions will usually not object.

PLANNING

When planning, designing and pasting up a publication, and choosing a printer, there are various points to bear in mind, and we will discuss these one by one below (⇨ also the chapter on Design and Paste-up, page 85, for full details about how to prepare your artwork).

SIZE OF PAPER AND IMAGE
Different presses take different sizes of paper, which is manufactured in standard sizes (see Glossary). Instant printers, in-plant printers, community printers and other small commercial printers generally use small offset presses taking A3 and/or A4 paper. Most larger commercial printers have A2 presses and even go up to A1. On a large press, it is always possible to print a smaller image than the paper size and trim down afterwards.

Margins. Because the machine grips the edge of the sheet of paper to take it through the machine, it cannot print right up to the edge, so you must leave a margin of 15 mm ($\frac{3}{8}$ in.) all round the printed image to allow for this. If you want the image to come close to the edge of the paper the margin can be trimmed off afterwards, which will cost you extra.

Bleeding. If a solid image or photo needs to come right to the edge of the paper (called 'bleeding') allow 3 mm ($\frac{1}{8}$ in.) of extra image on the paste-up. This is then trimmed back after printing.

SIZE OF PRESS AND QUALITY OF IMAGE
Large presses (A2 and A1) can print to a high standard whatever is on the plate, as long as a good paste-up is provided. There is often difficulty in printing large areas of solid colour and photographs on small offset presses (A4 and A3): better inking is required than the limited number of ink rollers on a small press can provide. This is why some instant printers are famous for their grey copies!

LENGTH OF RUN
In choosing a printer it is important to decide how many prints you need. The length of the run will affect not only the cost of the whole job but also the price of each copy (the 'unit price').

A leaflet with a run of 100 produced by young people on a further-education course (A4)

A community newspaper with a run of 1,200 (A3)

A fortnightly magazine with a run of 14,000 printed by web offset (A4)

Making the plates, putting them on the press and setting up the machine to print, are a major part of the costs involved in offset – once the machine is set up and running, the more prints are made the cheaper each print becomes. When getting quotes from printers ask for a 'run-on' price (that is, the cost of an extra 1,000 prints) – it may be worth printing more copies and so being able to sell them at a lower price.

Short runs are in general less economic, but the price will depend on the particular printer; instant printers, community printers and in-plant print-shops are usually more geared to producing short runs of simple jobs. The minimum run by an instant printer is likely to be fifty or 100; by a community printer 200 to 300. Larger commercial printers, which use metal plates rather than plastic or paper, cannot offer short runs as cheaply as instant printers – the minimum run is usually 500.

Larger commercial printers are as a rule cheaper for long runs than small printers. Printers using 'web-fed' presses – large machines fed by a reel of paper rather than separate sheets – can be much cheaper for very long runs (usually, anything over 5,000). This type of machine can also use print with cheaper paper such as newsprint, which helps to keep down costs, although it may not give such good-quality prints.

PLATES AND QUALITY OF IMAGE

The main factor affecting the quality of a printed image (provided you supply a good clean paste-up) is the type of printing plate used. Different plates also affect the cost, speed and size of print.

Plastic and paper plates. Instant printers are quick and cheap, mainly because they almost all employ quick plate-making methods using plastic or paper plates rather than metal ones. These plates are designed for short runs and not for high-quality work. They are made by a photocopying method, which usually means that it is not possible to enlarge or reduce from the paste-up as you can with metal plates. Because there is no negative stage, it is not very easy to remove shadows and unwanted marks. This problem can be avoided by making the paste-up as flat as possible in solid black on white. If necessary you can get a line print made (⇨ page 97) and then make the plate from this. For the best reproduction, photos should be screened with 65 or 85 dots to the inch (⇨ page 98). A more finely screened photo is likely to reproduce badly because the small dots which give the fine details may be lost.

Plastic and paper plates are available in A4 and A3 and can be used for runs up to 5,000. The two most common plate-making systems of this type are called 'photodirect' plates (PD) and 'chemical' or 'diffusion transfer' plates (CT or DT).

Metal plates. To produce a more durable plate, a negative photograph of the paste-up is made on a process camera and then exposed on to a presensitized metal plate with ultraviolet light. At the negative stage the printer can modify the image in various ways: screen photos into halftone dots; reverse out parts of the image; reduce or enlarge; paint out unwanted shadows or marks; or he can remove part of the image by painting it out with opaque paint.

Metal plates can be used for runs of many thousands and can reproduce finely screened photos and other fine details to a high standard. They are used for most larger commercial jobs but are more expensive than plastic or paper plates. If you need to use metal plates a smaller printer can get them made by a specialist plate-maker, although this will mean that the job will take a bit longer.

Direct-image plates. Direct-image plates are paper plates on which the image is drawn or typed directly. They are used for short runs only – some types can print no more than a few

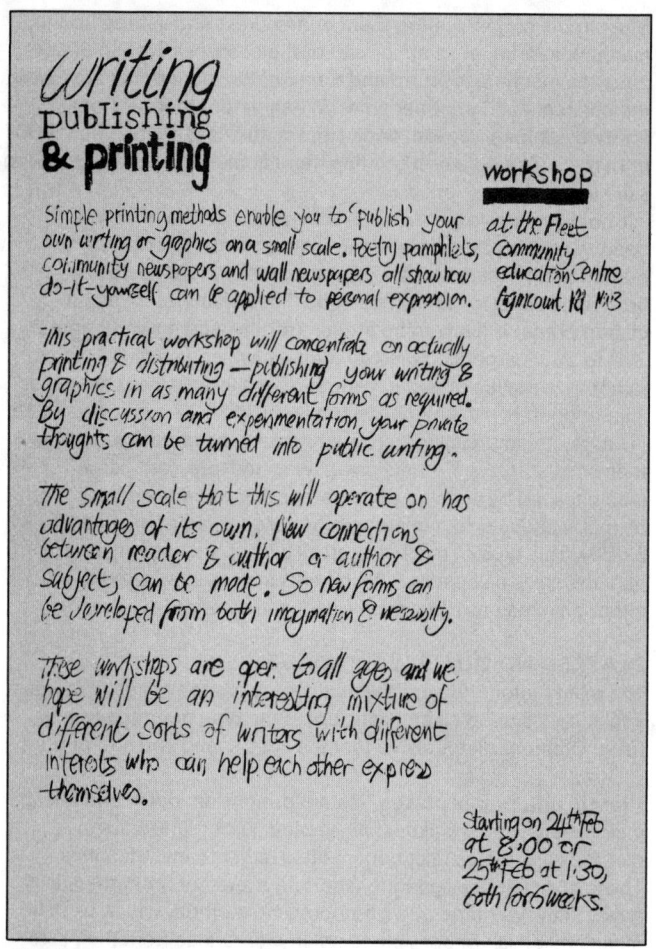

Leaflet printed from a direct-image plate on to which the information was handwritten (A4)

hundred sheets but most can produce up to 1,000. Direct-image plates are common in offices and in-plant printing where they are used to print typed office material on small 'offset duplicators'. They are cheap, quick and simple to use. The only problem is to find a printer who is happy to use them; you may have to go to a community printer or an in-plant print-shop.

How to make a direct-image plate
The paper of the plate is covered with a special surface that will accept a greasy image. You draw or type directly on the plate using a greasy ink or other water-resistant material. The plate is then wiped with 'etch', a special solution which makes the non-image area resistant to ink.

Special inks and pens are made for drawing on to direct-image plates and special typewriter ribbons for typing on them. But it is also possible to use wax crayon, some types of felt-tip pen, normal carbon typewriter ribbons, or carbon paper. The best approach is to experiment to see what works best – but be wary: even a greasy finger will make a mark on the plate, so do not touch the surface when drawing. Before you start to draw or type make sure you know exactly what you are going to do, and draw out a rough if necessary. Corrections are difficult, so be careful. If you do make bad mistakes, start again on a new plate. Do not press too hard when drawing as this makes it difficult to get a good print.

PAPER
Paper comes in a vast range of weights and thicknesses, colours, finishes and sizes. Of course, the price varies too! The printer can show you samples of the paper he stocks and what can be ordered. Tell him the colour and quality you want and what you can afford to spend, and follow his advice.

Weight. Paper weight is measured in grams per square metre (g/m^2) or pounds per ream. Instant printers tend to use 70 g/m^2 (45 lb) for single-sided work and 85 g/m^2 (60 lb) for double-sided work, and this is a good guide to follow for most jobs. Heavier papers will cost more and can increase postage costs considerably.

Finish. Paper is made with many different finishes. A glossy finish is more expensive than matt. The surface of the paper affects the reproduction of photographs; glossy art paper takes the fine details of a screened photo better than a cheap rough paper like newsprint.

PHOTOGRAPHS
Offset reproduces photographs well, although the quality will depend on the size of the press, as we have noted above. A photo printed by offset must first be converted into halftone dots or line.

Halftone. Photographs are converted into pictures made up of halftone dots by being rephotographed through a dot screen on a process camera (⇨ page 98). This turns the grey tones of the photo into a pattern of minute black and white dots which gives the illusion of tones. For high quality the screening is best done when the negative is made by the printer or plate-maker, but you can get a screened print made and then paste it down directly on to the artwork (⇨ page 105). You usually save a little by doing this. It also gives you more control: you can adapt the image (cut out the background etc.) and position it exactly where you want it on the artwork. (This could also be done at the negative stage but you would have to pay the plate-maker for doing it.)

The number of dots per inch you need will be determined by the type of plate the printer will be using, the type of paper that will be printed on, and the quality you require – the finer the dot, the more detail is retained.

Dot size	Plate	Paper
150	Use metal plate	On glossy paper (usually)
100	Use metal plate	On any grade paper
85	Use metal plate	On any grade paper
65	Use plastic or paper plate	On any grade or newsprint

Line. Photographs or illustrations can be converted into line. This turns all the greys into either black or white, making a strong bold image (⇨ page 97).

COLOUR PRINTING
Offset inks are made in a wide variety of colours, and if necessary the printer will mix up a special colour (but this costs extra). Coloured ink is always more expensive to use than black because the printer has to clean down the machine before and after using your colour. If you want to use colour but have a limited budget remember that it may be cheaper to print with black on coloured paper than to print directly with colour or use a second colour.

More than one colour. You can design your job so that it is printed in more than one colour ('Design and Paste-up', ⇨ page 85). A separate plate is made for each colour. The plates are then printed one at a time in their respective coloured inks on to the paper. The registration on most larger offset presses is very accurate but some small presses (A3 and A4, cannot register very accurately. If in doubt check with the printer and if necessary design your job so that it does not have to rely on precise registration (⇨ page 105).

Coloured inks are affected by the kind of paper used. Colours printed on glossy white paper (as in most sample books) look brighter than when printed on cheap-quality papers.

Left: Cover of adult-education-centre magazine, showing the use of a screened photoprint with the background cut away (A4)

Cover of poetry book using a bold line photo (21 × 11·5 cm)

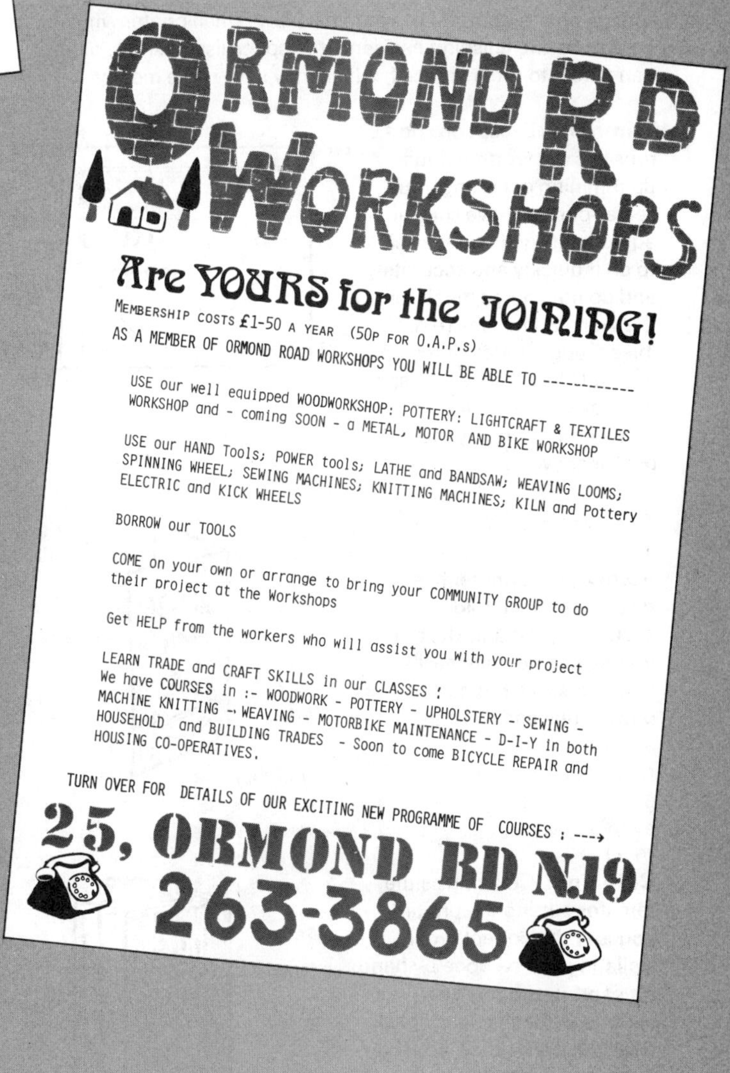

Left: Two-colour cover of two-monthly magazine (28 × 20 cm)

Above: Two-colour leaflet to advertise craft centre (A4)

Full colour. 'Full' or naturalistic colour is produced by combining four 'process' colours: yellow, cyan (a light greeny-blue), magenta (a bluey-red) and black. The full-colour original is broken down or separated into these four colours photographically to make four separate screen negatives. Four plates are then made from these negatives. Each colour has to be printed in perfect register on top of the next to maintain the illusion of full colour. If you look closely at a large advertising hoarding you can easily see the separate coloured dots.

Making colour separations is a skilled job and an extremely expensive process, used for very long runs of items like brochures, book jackets etc. Modern computer-controlled machines are now being used to automate the process and this may eventually make it cheaper.

FINISHING

'Finishing' is the general term for all the jobs that have to be done to printed pages to make them into a finished publication. It usually includes some or all of the following:
- Trimming.
- Folding.
- Collating.
- Binding.

With a run of under 2,000 you can do some of these jobs yourself – they may be very boring but with willing helpers they can be done fairly quickly! If finishing is done by the printer it may form a large proportion of your final bill. A small printer who cannot do the finishing can send it to specialist finishers, or you can take it to them yourself, which may save some money.

Trimming. Except on short runs it is not worth trying to do trimming or cutting yourself as printers have special equipment (guillotines etc.) to do it quickly and accurately, and do not charge much for the work. Some jobs (e.g. 'bleed' edges) are printed on a slightly larger sheet of paper and have to be trimmed down to the correct size by the printer anyway.

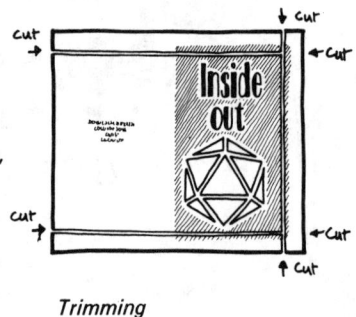

Trimming

Folding. The printer has machines which fold accurately and quickly, but folding can be done easily by hand for short-run newspapers, newsletters and so forth.

Folding

Collating. 'Collating' is the term for putting the printed pages in the correct order. Collating can be done by hand for short runs but is very tedious. A mechanical 'collator' does the job quickly; you may be able to use one in your office or at a local resource centre or community printer.

Collating

Binding. *Stapling*, called 'stitching' in the trade, is the cheapest binding method. It is also easy to do yourself if you have access to a large stapler. The large staplers that printers use can staple through a maximum of one hundred sheets (beyond which a book will not fold shut properly). For thin books the stapling can be done through the centre fold (called 'saddle-stitching'). For thicker books it has to be done down the side (called 'side-stitching'); this has the disadvantage that the book will not open flat.

Saddle-stitching   *Side-stitching*

A variety of mechanical binding systems using *metal or plastic coils* are now available for short-run publications. A series of holes or slots are made down the edge of the pages and the cover, and they are then bound together by a metal or plastic coil. You can do this kind of binding yourself if you can get hold of the equipment from an office, school or college. It produces a publication that opens up flat, and additional pages can go in at a later date if necessary. Mechanical binding can also be done for you by the finishers.

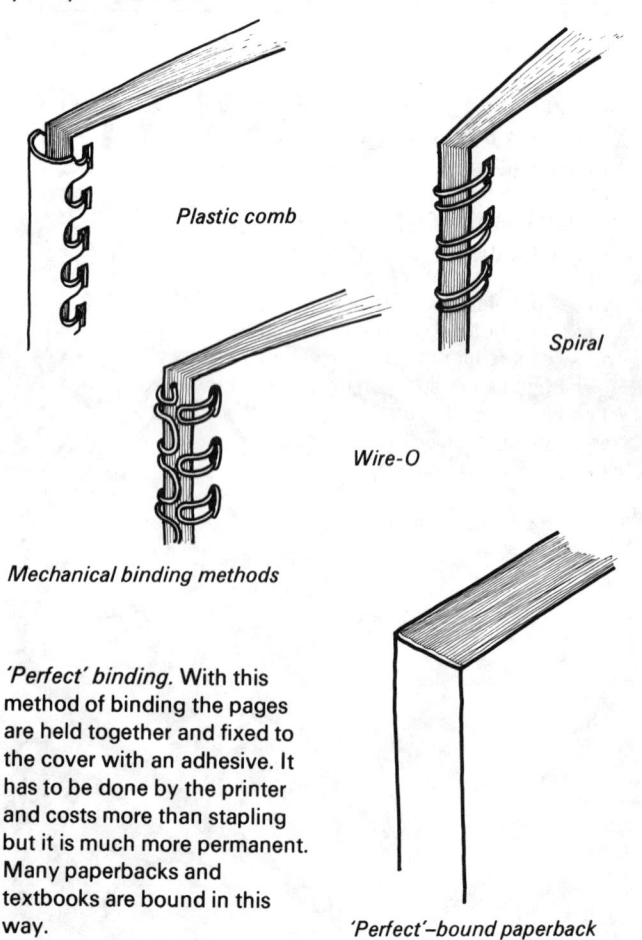

Plastic comb

Spiral

Wire-O

Mechanical binding methods

'Perfect' binding. With this method of binding the pages are held together and fixed to the cover with an adhesive. It has to be done by the printer and costs more than stapling but it is much more permanent. Many paperbacks and textbooks are bound in this way.

'Perfect'-bound paperback

KEEPING A SMALL JOB SIMPLE, QUICK AND CHEAP

Here are a few tips which may help you keep short-run jobs cheap and simple:
- *Printer.* Find a printer who specializes in the kind of job you want done.
- *Size.* Design your job to a paper size that small offset presses take (A3, A4, 10 × 16 in., 11¾ × 17 in., etc.; ⇨ Glossary for paper sizes).
- *Plates.* For runs up to 500 look for a printer who uses plastic or paper plates. These are cheap and simple to use, although they may not produce top-quality prints.
- *Size of paste-up.* Make your paste-up the same size as the finished print, in view of the fact that the printer using simple plate-making systems may not be able to reduce or enlarge.
- *Quality of paste-up.* Provide the printer with a clean flat paste-up with strong blacks on white paper; this will give you the best results if plastic or paper plates are used.
- *Making a photo-print.* To improve the quality of your paste-up so that plastic or paper plates can be used, get a photo-print made. If the paste-up needs enlarging or reducing this can be done on a photo-print too (⇨ page 102).
- *Colour.* Most small printers keep their presses inked up with black and charge extra to clean it off and use another colour – so it's cheaper and quicker to use black. If you want some colour consider using coloured paper instead.
- *Finishing.* To keep costs down, if the run is short see if you can arrange to do your own finishing.

CHECK-LIST

Before approaching printers for quotes you must plan exactly what you want. By now you should have thought about the following points:
- ☐ Size of finished job.
- ☐ Length of run.
- ☐ Number of pages/printed both sides?
- ☐ Choice of paper (weight, colour and finish).
- ☐ Number of colours/choice of colours.
- ☐ Number of photos to be screened if the printer is doing it.
- ☐ Any reversing out or tone to be carried out by the printer?
- ☐ Finishing to be done (trimming, folding, binding).
- ☐ When will the job be ready?
- ☐ Do you want it delivered?
- ☐ When do you have to pay?

DEALING WITH THE PRINTER

While you are planning your job you may need to ask for a quote to see what kind of result you can afford. What you are given may only be an estimate, since some printers will want to see the paste-up before giving a definite price.

Once you have decided what you need, obtain quotes from two or three printers, giving them the information on your check-list, some of which will need to be discussed with them. (You can do this on the phone or call in on them.) Decide which printer you are going to use, bearing in mind what type of printer they are and the quality of their work, as well as the price they have quoted.

If possible, take your paste-up to the printers yourself. By doing so, you are able both to avoid the risk of loss or damage in the post and to show it to the printers to check that everything is in order. Make sure that they have all the necessary details (see your check-list); for a large job it may be best to write all the details in a covering letter; for a small job attach the details to the artwork or write them on the back.

If you find printers who are helpful and do a good job, use them regularly – you'll get to know them and may be able to learn a lot from them. Printers are naturally enough the best people to teach you about print!

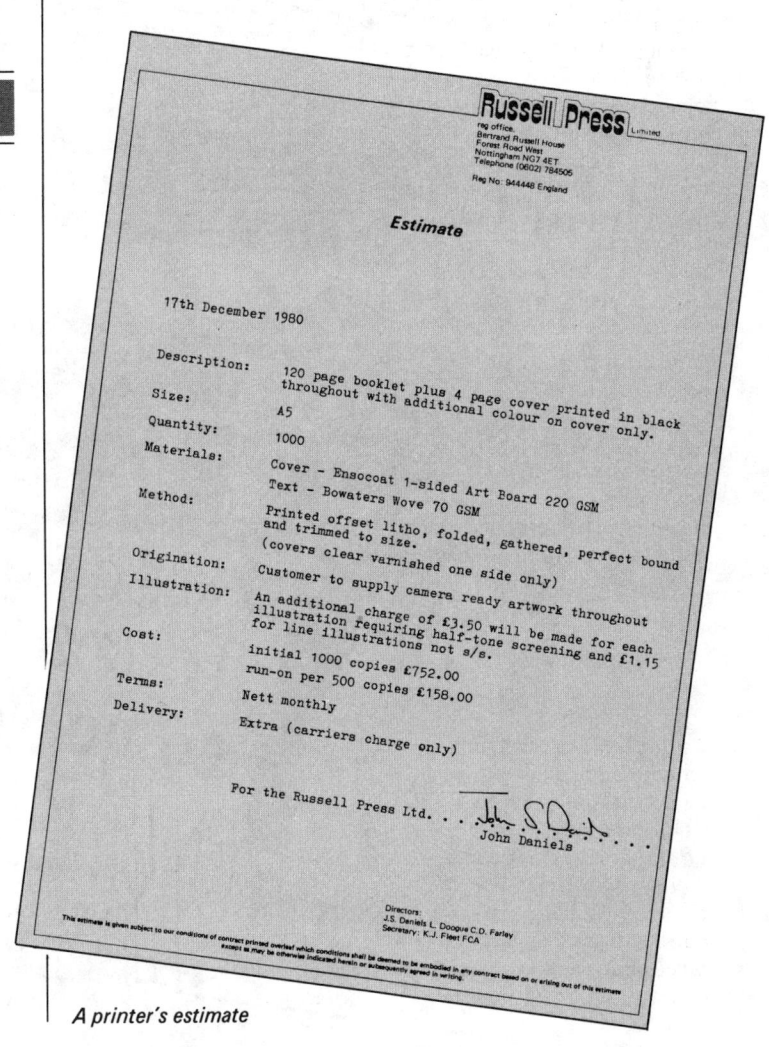

A printer's estimate

Poster produced at short notice to advertise public meetings, using a collage of photos and lettering (A3)

Other Printing Methods

Photocopying 71
Access 71
When to use a photocopier 71
Plain-paper copiers 71
Coated-paper copiers 72
Full-colour copying 72
Copying on to other materials 72
Using a photocopier 73
Renting or buying your own copier 73

Relief printing 74
Cutting a block 74
Making a collage block 75
Inks 75
Inking the block 76
Printing with the block 77
Cleaning up 77

Rubber stamps 78
Making your own stamp 78
Ready-made rubber stamps 78
Getting rubber stamps made to order 78

Letterpress 79
Hand letterpress 79
Commercial letterpress 80

Spirit duplicating 82
How it works 82
Colour 82
Making the master by hand 83
Using a spirit duplicator 83

PHOTOCOPYING

Photocopying is the general term for copying methods which use the electrostatic process. Invented in 1938, this is a comparatively modern process which has been developed and improved a lot in the last few years. Photocopying is now increasingly cheap, flexible and of good quality and it has become one of the most 'instant' and accessible copying methods.

There is a vast range of machines available. Most copy on to A4 or legal-size paper and some can copy on to A3 size. Recent developments have led to machines that can reduce to various sizes, copy on to both sides of the paper, print on to different types of material (acetate, for example), and make full-colour copies. Some have built-in collators.

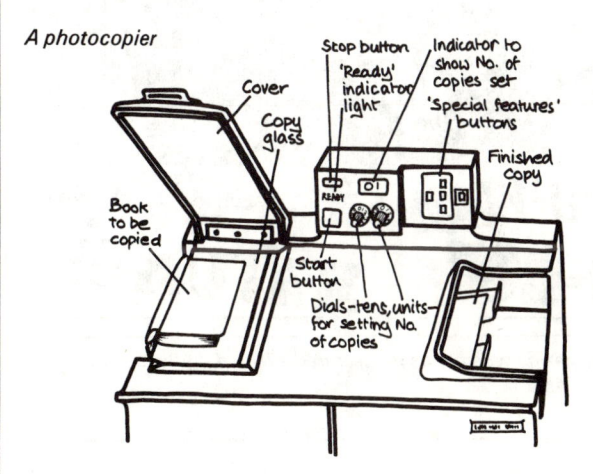

A photocopier

One of the drawbacks of photocopying is the quality of the print. Blacks may be very grey, solid blacks are difficult to achieve and the white background may look dirty. However, this to a large extent depends on the kind of machine used and whether it is kept clean and well maintained, and machines are now being developed which give better solids and cleaner copies.

ACCESS
Photocopiers are found in 'copy shops', instant printers, offices, schools, libraries and resource centres. Prices vary greatly – from 10p (10¢ usually, in the U.S.A.) per copy at a copy shop down to 4p (5¢) per copy at a resource centre. Sometimes you are allowed to operate the machine yourself – in fact many machines in public places are coin-operated. In copy shops the copying is done for you. At copy shops and other places where the copies are made for you, quality may be better because larger machines are used, and these places are often able to offer a range of coloured papers and other services like collating.

WHEN TO USE A PHOTOCOPIER
Photocopying is mainly useful for short runs of under fifty copies: for longer runs other methods like stencil duplicating and offset-litho may be cheaper and give better quality. But photocopying is in frequent use because it is so quick and time-saving. (Photocopying is obviously a very convenient method if you have access to a machine that you do not have to pay to use.)

PLAIN-PAPER COPIERS
The plain-paper system uses the electrostatic transfer process. Because the image is copied first on to a drum and then transferred on to the paper, any good-quality plain (uncoated) paper can be used. This means you can copy on to white paper,

Music magazine which puts the textured effect of photocopying to good use (A5)

Cover of newsletter using the interesting texture produced by the copier (A5)

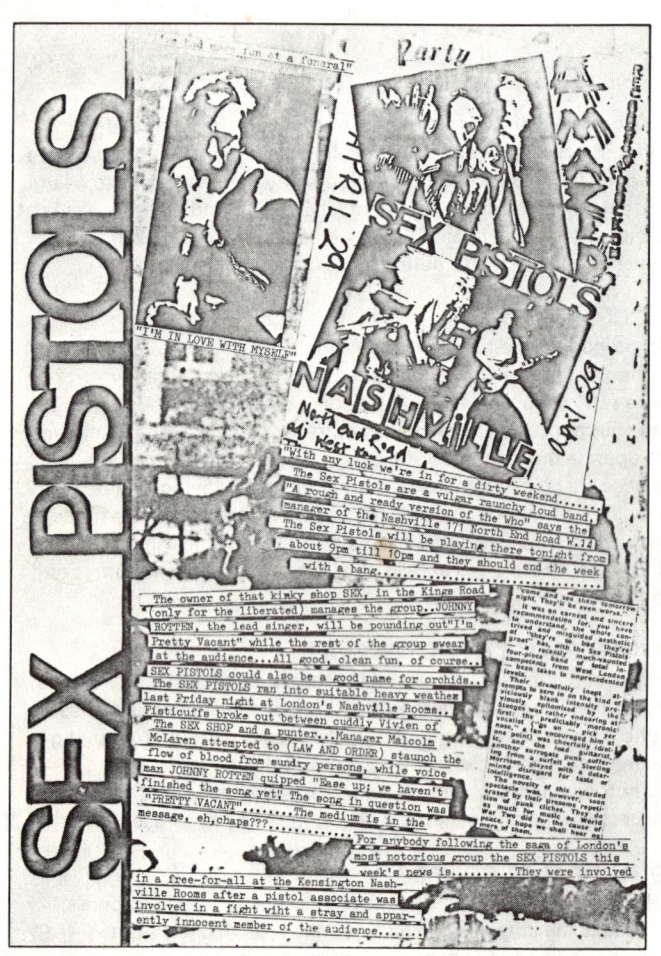

Page from a punk 'fanzine' uses a collage of photos and text (A4)

coloured paper, pre-printed paper (such as letterheads) or on to both sides of the paper.

The quality of the print depends on the machine and how well maintained it is, but plain-paper copiers tend to make greyer solids and dirtier copies than coated-paper copiers.

COATED-PAPER COPIERS

The coated-paper method uses a direct electrostatic system. The image is made straight on to the paper, which has a special light-sensitive coating such as zinc oxide. Machines based on this system are less common and more expensive to use than plain-paper copiers and in some types the paper has to be developed in liquid chemicals.

Better contrast and deeper tones than from a plain-paper copier are obtainable on these machines, many of which also have contrast controls, enabling you to achieve a better-quality image. Sometimes the paper is cut from a roll so you can set the machine to cut copies to the required size.

FULL-COLOUR COPYING

Some photocopiers can produce a full-colour print from coloured artwork, coloured photographs or full-colour slides. This is still a very expensive process (at the time of going to press it cost between 70p/$1 and £1/$1·50 per copy), but is as yet the only way to produce short runs in full colour. As the technology develops and the machines become more common this process may become cheaper.

COPYING ON TO OTHER MATERIALS

It is possible to make photocopies on materials other than paper (acetate, for example, to make transparencies for overhead projectors), but you may have to hunt around to find this service. Copies can also be made on heat-transfer paper, from which the image is transferred to fabric by ironing, but this process is not yet widely available.

USING A PHOTOCOPIER

Photocopying is a simple push-button method. You place the original face down on the glass, set the dial for the number of copies you want and push the button. Because the machine is made so that only simple adjustments need to be made by the operator, and because there are so many different models, we feel that rather than tell you how to operate a particular machine, the best advice we can give is in the form of tips on how to get better-quality copies.

- Make one copy to check the positioning and quality of the print before you go on to do a long run of copies.
- Make your original without large solid areas.
- You can improve the quality of photos or large solid areas by laying a white-dot screen on top of the original. (The screen can be obtained from the manufacturers.) This breaks up the black areas so that they reproduce better. You could make your own dot screen by sticking down a sheet of white instant tone on a piece of clear acetate and then laying this on top of the original.

Placing a white-dot screen over a piece of artwork

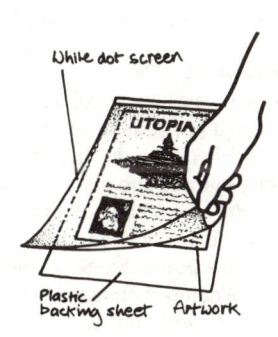

White-dot screen in use to improve the reproduction of a photo (enlarged detail)

- Try using photos from magazines, or newspapers, which have already been screened. Simply stick them down on the original.
- If you want a photo to reproduce to a high quality it may be worth getting a screened photo-print made (⇨ page 98).
- If you get grey blacks and a dirty background the machine may need more 'toner' added by the operator. (If the machine belongs to you, keep it well maintained and cleaned.) If necessary, adjust the contrast control, or push the 'faint original' button to get darker copies.
- If you are using a pasted-up original and paste-up edges can be seen, try taking the first copy as your original, painting out the

> **Raising Money from Government**
> LOCAL AUTHORITIES
> Grants from local authorities often provide many groups with their basic running costs. The method of application varies from authority to authority depending on the departmental and committee structure. Some

Paste-up edges being picked up by the copier (detail from a page of a fund-raising pamphlet)

marks with water-based white paint. (Do not use spirit-based white as this will dissolve the black in the copy, turning it grey!) Then use the copy as the original to make the rest.

- You can make reductions on a reducing photocopier instead of getting a photo-print made, which is much more expensive... but make sure you use a machine that gives a good-quality strong black (not grey) copy.
- To avoid grey shadows around the edge of the paper check that the original is in the correct position on the glass. Or try placing a large sheet of white paper on the back of the original once it is laid in position on the glass of the copier. This is very helpful when reducing, or if you are copying something smaller than A4.
- If you are using a machine that reduces, you can tape two A4 pages together and copy them on to one A4 sheet. This can then be folded easily and quickly into an A5 leaflet (⇨ page 89).

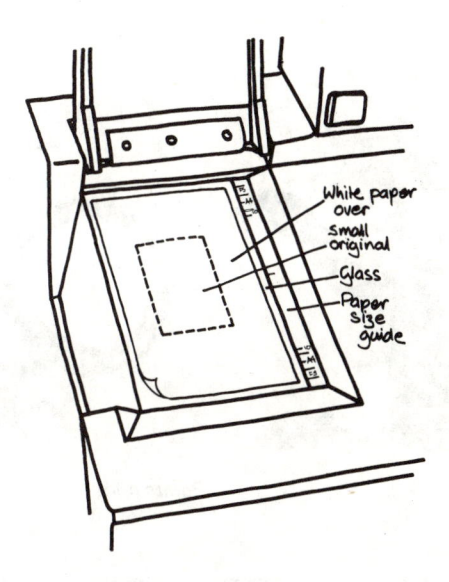

Placing a large sheet of paper on the back of the original

RENTING OR BUYING YOUR OWN COPIER

If you do enough photocopying it may be worth renting a copier (or sharing one with another organization). Because there is usually a 'minimum copy' billing system it is not worth renting a copier unless you do enough copies in a month to cover the bill. For example, in the U.K. if you rent a copier which makes A4 and A3 copies, the monthly bill is £75 (plus VAT) of which £28 is the rental charge and the remaining £47 is the 'minimum copy' charge – you pay this even if you have done copying worth less than £47. If you do more copying than the minimum, you pay pro rata for each copy above the minimum copy charge. Similarly, in the

U.S.A. the monthly bill for renting a Xerox 3107 will be $200, which gives you a minimum of 1,100 copies. Additional copies are charged at approximately ½¢ each. So you need to make a minimum number of copies per month in order to be able to cover your costs while charging a reasonable price per copy.

It pays to work out what your average volume of work is monthly and to decide on the quality you require and how rapidly you want to make copies – and then to talk to the reps from the various companies to see which model and brand suits your needs best. If you are leasing or renting a machine the monthly billing system covers repairs and servicing, but if you buy a machine it is worth taking out a service contract with the manufacturers, paying a monthly amount to cover any necessary repairs.

If you are going to let other groups use the machine, work out your charges so that you cover all your costs, including wastage and your service contract, if you have one. Photocopiers are too complex to repair yourself, so remember that your machine may well be out of action for short periods while you wait for the engineer. You can go on a free operators' course run by the manufacturers, to learn to sort out minor problems like freeing jammed paper.

RELIEF PRINTING

Relief printing is a very straightforward process: a raised surface is covered with ink and then pressed on to paper to make a print. The simplest form of relief print can be made using an actual object, such as your hand or a cotton reel, but most relief printing is from a 'block', in which a design is cut in reverse into a suitable material. The parts that are cut away do not print, the image being made by the raised areas.

Relief printing comes in handy for many things: cards, logos, rubber stamps (⇨ page 78), printing on fabrics, illustrations for letterpress printing (⇨ page 79) or for adding a spot of colour to black-and-white copies.

In this section we give a simple outline of relief printing so that you can understand how it works and see whether it is suitable for your particular needs. If you want to pursue the subject in more depth we suggest you read one of the many good books available on the subject (⇨ page 109) and experiment to find the best method for what you want to print.

CUTTING A BLOCK

Various materials can be used to make a block, depending on the kind of image you require. Quite soft materials like potatoes can be used for simple crude shapes, but for designs with fine details a harder material like linoleum or wood is needed.

Simple shapes can be cut into a potato with a sharp knife, but the edges of the shape tend to go soft quite quickly once you start printing, and a potato block will not, of course, keep for future use. More complicated shapes can be cut into balsa wood or polystyrene (styrofoam) with a sharp knife. Blocks made in this

Relief prints

Hand print  *Potato print*

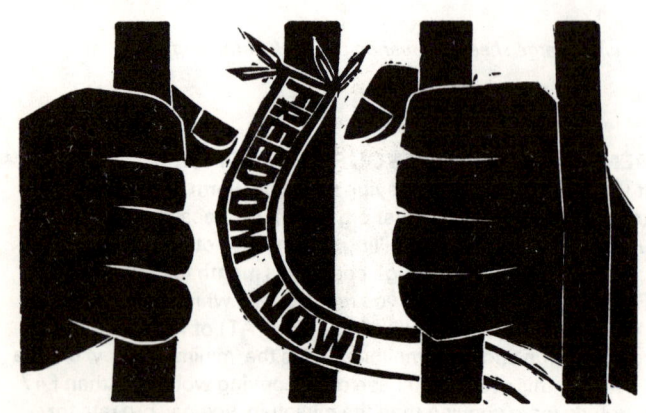

Linocut image on fund-raising card (11 × 15·5 cm)

Woodcut reproduced as a poster by offset (51 × 37 cm)

RELIEF PRINTING 75

way can be cleaned and stored for later use. For harder materials like lino, or wood, you need special lino- or wood-cutting tools, which you can buy at craft shops.

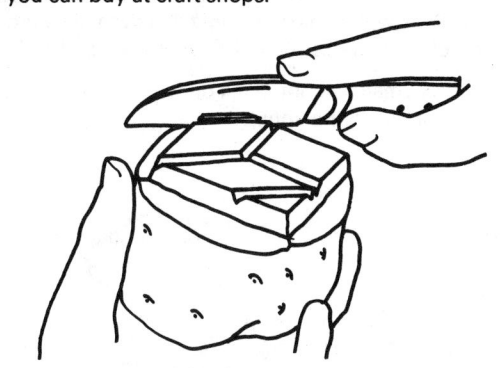

Cutting a potato with a knife

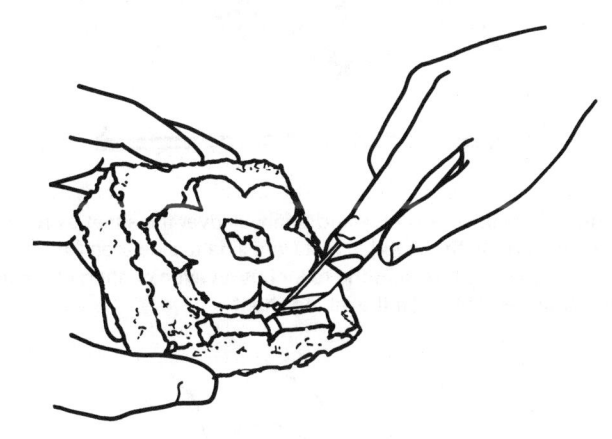

Cutting polystyrene with a craft knife

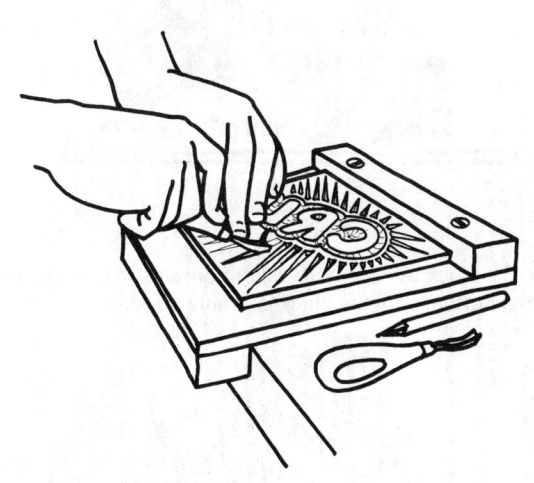

Cutting a lino block with a linocutting tool

Remember that the design must be cut *in reverse*, otherwise it will print back to front. This may not matter with a picture, but it is obviously important with lettering. Here is one easy way to reverse your design.

1. Draw out your design in soft pencil on tracing paper.

2. Transfer the design in reverse on to the block by laying the tracing upside-down on top of the block and going over the design again with a hard pencil.

3. This will leave a faint outline of the design on the top of the block.

MAKING A COLLAGE BLOCK

If it suits your design better you can make a block by sticking things down on to cardboard or wood. The shapes can be cut out of cardboard, fabric, corrugated cardboard etc., or can be patterns made with string or other objects. Whatever you use to make the relief pattern, make sure that it has a uniform height or the block will be difficult to ink and print.

Collage block made by sticking cardboard shapes and string on to cardboard

Print taken from the collage block

INKS

For printing small simple blocks made from such things as potatoes almost any thick poster paint will do. For more detailed lino or wood blocks you will only get a good-quality print if you use block-printing inks. These can be bought at art and craft shops and are available as water-based or oil-based inks. Oil-based inks give better-quality prints but take longer to dry and are much messier to clean up, because you have to use spirits. Block fabric-printing inks are also available (oil-based or water-based) if you want to print onto fabrics. All these inks come in a large basic range of colours which can be easily mixed to give whatever colour you require. (Don't try to mix oil-based with water-based inks . . . they won't!)

A print taken from a card block shows the uneven printing quality

Inking the block

Stamp printing. Smaller blocks (potatoes, cardboard, rubber stamps) can be inked simply by pressing the block into an ink-pad or a sponge soaked with ink or thick poster paint. Blocks inked in this way tend to make prints with a slightly uneven look; this can give an attractive improvised look.

Ink pad made from a sponge and tin lid

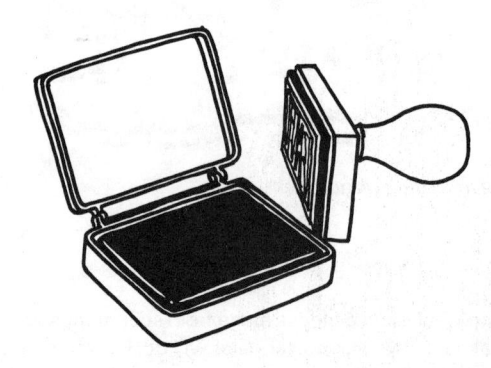

Bought ink pad

Painting the block. Simple blocks like potatoes can be inked by painting the ink or paint on with a large brush. This is a good idea if you are only making a few prints.

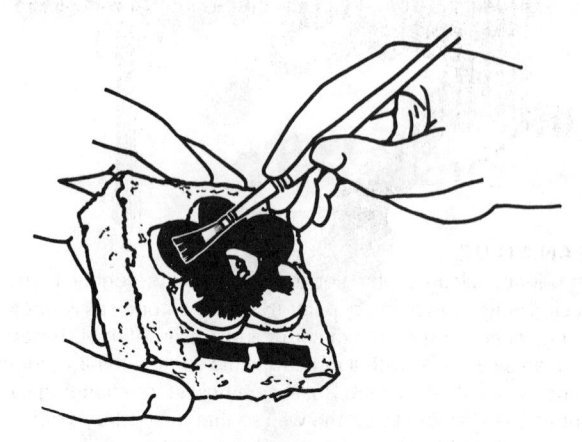

Applying paint to a polystyrene block with a paintbrush

Using a roller. Larger blocks and those with fine details will not ink well enough with either of the above methods. Instead, ink is applied to the block with a roller, giving a thin even covering. Rubber rollers of various sizes can be bought at art and craft shops.

1. Squeeze a very small amount of ink on to a smooth surface (a sheet of glass, Formica or Perspex).

2. Spread the ink out evenly and thinly all over the sheet by rolling back and forth with the roller, first from side to side then from top to bottom. Then make sure the roller has an even coating of ink by rolling back and forth on the inking sheet.

3. Ink up the block by rolling the roller from side to side and then from top to bottom until the block is evenly inked.

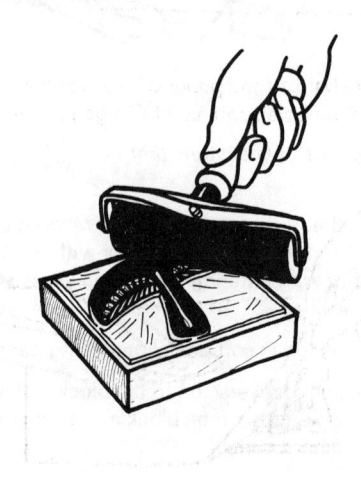

Printing with the block
If you are stamp printing, press the inked-up block firmly and evenly down on to the paper. Re-ink the block after each print.

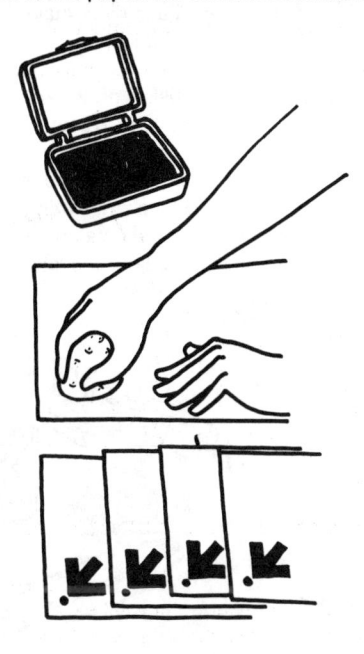

One problem is that large or detailed blocks do not print well by the stamp method. To get round this, lay a sheet of printing paper on top of the inked-up block and apply pressure to the back of the paper by one of the following methods:

1. Rolling a clean hard roller firmly back and forth across the paper, *or*

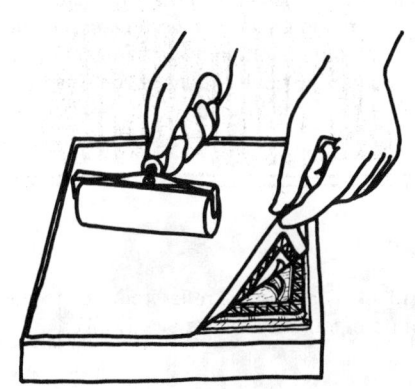

2. Burnishing the back of the paper with a wooden spoon, going round and round in circles until you have gone over all the print area, *or*

3. Putting a board on top of the paper and then standing on it, or rolling a garden roller across it (this is a good method for large prints), *or*

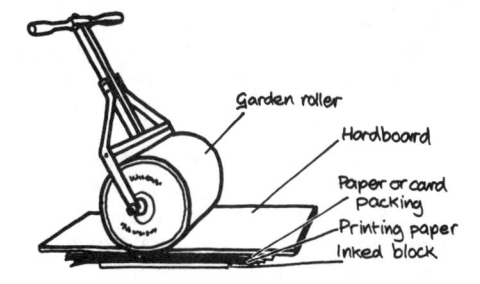

4. Using a printing press. To keep all the fine detail and to get evenly inked prints from very detailed blocks like woodcuts and lino cuts, the strong even pressure of a proper relief printing press may be necessary. An Albion or Colombian proofing press may be available at a local school, college, print-making workshop or adult-education institute (some of these places may also run evening classes in block printing).

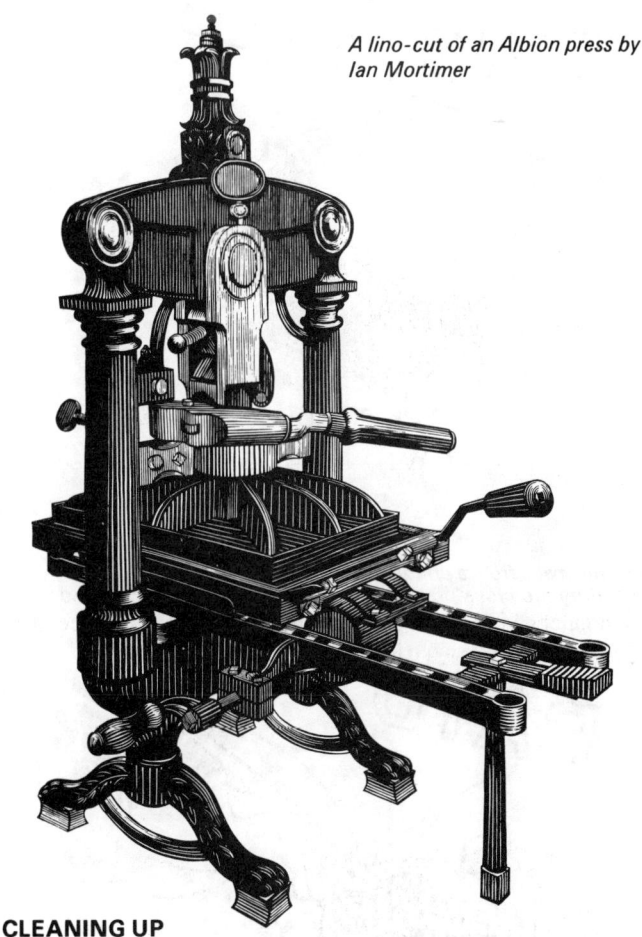

A lino-cut of an Albion press by Ian Mortimer

CLEANING UP
After printing, clean up the blocks and rollers thoroughly. If you are using water-based ink or paint this can be done easily under running water using a cloth or scrubbing-brush. With oil-based inks clean everything with a rag and white spirit or mineral spirits until no traces of ink remain. Rollers are best stored hanging by their handles from hooks on the wall so that the rubber is not damaged. Clean blocks can be stored and used again.

RUBBER STAMPS

One of the most common types of relief blocks for small prints is a rubber stamp. Such stamps are in use in offices for making common additions to forms, letters, envelopes and other documents. A rubber stamp with your name and address can be very useful for letterheads, invoices, envelopes, cards and the like. For small quantities it is much cheaper to do this than to obtain commercially printed stationery.

The main drawback of rubber stamps is that they do not make a very even print. This is particularly noticeable if you print areas larger than 8 cm (3 in.), or solid areas. But you can print on to anything, and you do not have to print more than the exact number that you need at the time. Because each print is made by hand, long runs are not very practical.

You can make rubber stamps from a kit, buy them ready-made, or get them made to order, depending on the kind of lettering or image you require and the size.

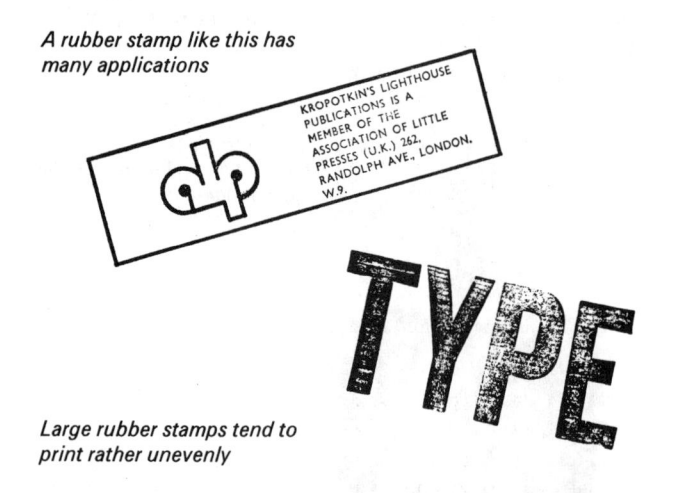

A rubber stamp like this has many applications

Large rubber stamps tend to print rather unevenly

Stamp made from a printing kit

Bought rubber stamps

MAKING YOUR OWN STAMP
Rubber stamps consisting of a couple of short lines of words can be made from rubber printing kits (such as John Bull or Speedball). These kits are quite cheap and can be bought at art shops, toy-shops and stationers. Each kit contains only one size and style of type, which may be a bit limiting.

Using a printing kit. The kit consists of: letters set in relief in blocks of rubber, pincers to set the type, a stamp into which the letters are slotted, and an ink-pad. Using the pincers or your fingers, slot the letters into the stamp in reverse order. Check that you have set the type correctly by making a print on rough paper – then use the stamp like a normal rubber stamp. When you have finished with the stamp you can take out the letters and store them until you need to make another stamp.

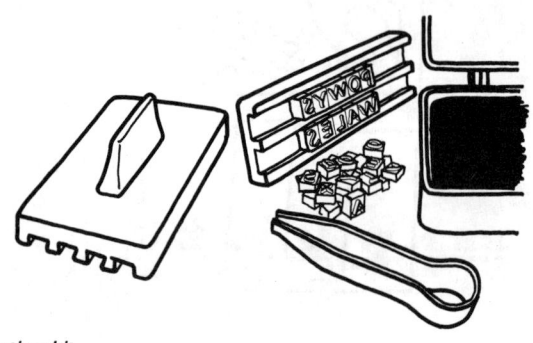

A printing kit

READY-MADE RUBBER STAMPS
Many stationers carry a range of rubber stamps for office work – for devices like dating and numbering machines, stamps saying 'paid', 'cancelled', 'urgent', 'rubbish', and sometimes with symbols like arrows and stars, which can be used in interesting ways. Rubber stamps with pictures of plants, flowers, animals and such-like can often be found in toy-shops, art and craft shops and gift shops. These are usually quite expensive, costing between £1 ($2) and £3 ($6) for a small stamp.

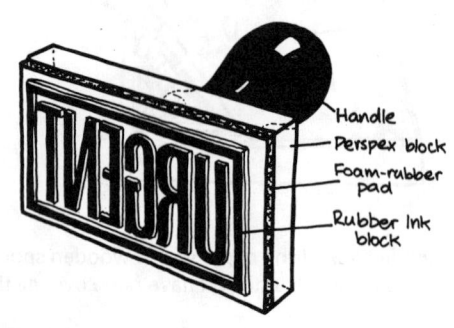

A rubber stamp

GETTING RUBBER STAMPS MADE TO ORDER
If you need a size or style of lettering that is not available in a kit, or if you want images other than words, you can get a rubber stamp made for you. Many larger stationers and some smaller ones are agents for rubber-stamp makers, or you can go to them direct. 'Rubber Stamp Makers' are listed in the yellow pages.

From the catalogue. The cheapest way to get a stamp made is to use one of the standard typefaces that the stamp-maker keeps in stock. The range will be shown in a catalogue or with samples.

LETTERPRESS

All you need to do is to make a rough drawing of the wording and how you want it arranged, and give details of the size of lettering and size of the stamp you want.

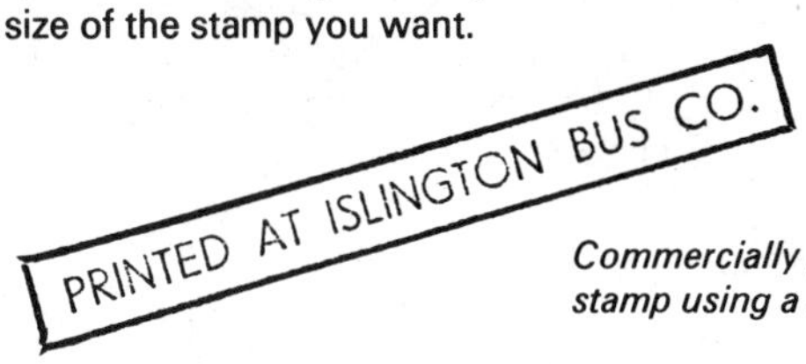

Commercially produced rubber stamp using a stock typeface

From a paste-up. If you want a typeface that the stamp maker does not stock, or other kinds of images, then you can get a stamp made photographically from your paste-up. Check with the stamp maker beforehand that they can do this (not all do), and ask what size you should make the artwork: it might need to be the same size. This process is expensive (£10 to £15/$20 to $30), but it is worth it if, by using the stamp, you are saving the expense of getting other things printed.

Commercially produced rubber stamp made photographically from artwork

Commercially produced rubber stamp made photographically from artwork using an image as well as lettering

LETTERPRESS

Letterpress is a very old form of printing, used in Europe since the fifteenth century. Movable relief letters made of metal called 'type' are arranged by hand into words, lines and pages. This is called 'hand composition'. The raised surface of the type is covered with ink and the print made from it. Many small printers and hobby printers still use hand composition, but most large commercial letterpress printers of newspapers, journals etc. use machine composition, where they have not yet joined the industry-wide shift away from machine composition to methods of photocomposition.

In this part of the section we deal briefly with hand letterpress: what it may be useful for, how to find or get access to the equipment, basically how it works, and how to find out more about using your particular press. The second part of the section tells you how to use a commercial letterpress printer.

Metal type in the composition 'stick' (hand-setting)

HAND LETTERPRESS

Access. Hand-letterpress equipment is not all that common, but there is still a surprising amount of it about, often unused and stored in attics or garages. Schools, colleges and adult education institutes occasionally have hand-letterpress equipment that you can use at evening classes. Second-hand equipment is frequently for sale cheaply through classified ads in magazines (such as *Exchange & Mart* in the U.K.) – you need only a small amount to get started, and can always buy more later.

Uses. Hand letterpress can be used to print small things like letterheads, cards, tickets etc. Some people become very involved in the process and go on to print high-quality publications, such as poetry books. This is very satisfying, but requires a great deal of time, patience and skill. Hand-setting and printing are time-consuming, so long texts and long runs are not advisable; beginners should find runs of fifty to 200 quite manageable.

Letterhead printed by letterpress (25·5 × 20 cm)

80 OTHER PRINTING METHODS

Illustrations. You can get good prints from lino or wood blocks with letterpress. This is a cheap way to make illustrations, headlines or other symbols. Whatever kind of block you use make sure that it is the same height as the type (usually 2·33 cm/0·918 in.); if necessary, mount it on a wooden block. The block itself is then put in position with the set type and the two are printed at the same time.

Hand-cut relief blocks are basically easiest; if you want to use drawings or photographs you will have to get a block made photographically by a photo-engraver (⇨ 'Process Engraving' in the yellow pages). Such blocks are comparatively expensive, but they may be cost-effective if you are printing a long run.

Presses. The two main types of machine you are likely to come across are the flat-bed and the small hand letterpress.

The flat-bed press. Whatever their size, all flat-bed presses are used in basically the same way. Set type is positioned on the flat bed of the press (large wooden type can be set directly on to the bed) and held in place by a fastening system (magnets, for example). The type is inked up with a roller, the paper placed on top of the type and the print made by pulling the impression roller across the paper. This is a slow process, thus only suitable for short runs – fifty prints takes about half an hour.

Some flat-bed presses are made specially for use with large wooden type for printing short runs of small signs and posters – some of these presses are quite small and fit on top of a table. These presses are sometimes used in schools and colleges to print posters.

The larger flat-bed presses are big, heavy, floor-standing machines originally made for use as proofing presses. They are most likely to be found in art schools and, sometimes, adult education institutes.

The small hand letterpress (e.g. Adana). With small hand letterpress the type, when set, is fastened together in a frame and clamped in position on the bed of the press. With the pull of a lever, ink rollers take ink from the ink disc and ink up the type, after which the paper is pressed up against the type to make the print.

The Adana is one of the best known and most widely used of these small 'platen' presses (a platen press is a press in which the print is made by pressing two flat surfaces together). It is small, easy to use and particularly useful for small items like cards, invitations, tickets, programmes, letterheads, and so on. It comes in a variety of models that can print on to paper sizes ranging from 13 × 20 cm (5 × 8 in.) up to A4 (8¼ × 11¾ in.).

Companies like Adana produce illustrated leaflets for all of their models describing all the stages in setting the type and making the prints. So if you manage to buy or borrow a machine write to the manufacturers for details on how to use that particular model.

COMMERCIAL LETTERPRESS

Letterpress is still used commercially, but by a diminishing section (about 40 per cent) of the printing industry, ranging from small jobbing printers (⇨ page 81) to very much larger companies printing newspapers and books. Letterpress is becoming less common and more expensive, giving way to offset-litho and photosetting, which tend to be cheaper for many jobs.

Uses. Letterpress may be cheaper for small jobs like invitations, or for jobs that are not suitable for offset-litho:
- You may need to use letterpress if you want to have high-quality justified typesetting in an unusual type face, e.g. for a book of poems produced to a high standard. (But a good-quality letterpress proof can be used to make artwork for offset-litho.)
- Large jobs made up entirely of text requiring good quality and long runs, such as a book, may be better in letterpress, so obtain quotes for comparison from both a letterpress and a litho printer.

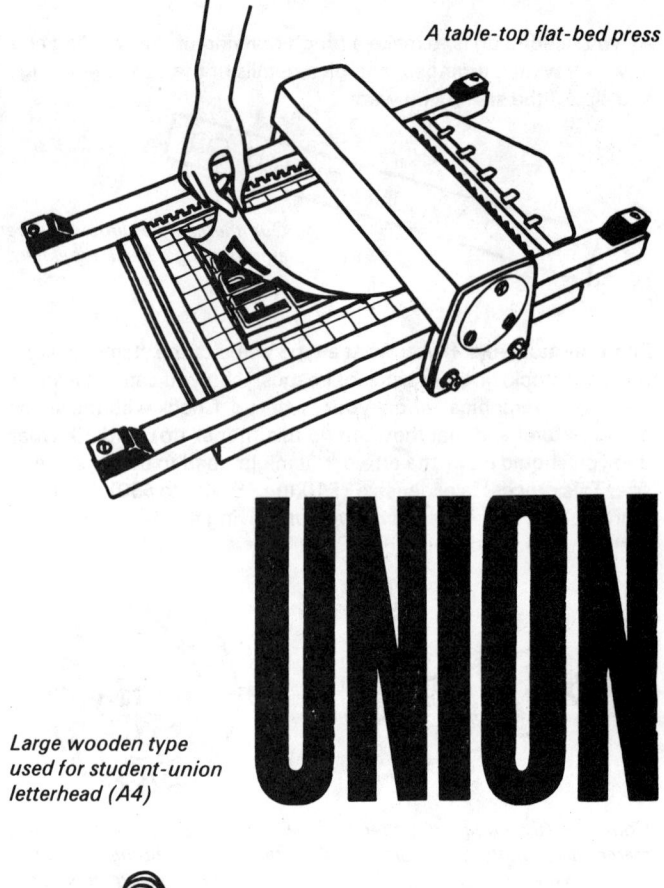

A table-top flat-bed press

Large wooden type used for student-union letterhead (A4)

Large proofing press with wooden type in position on flat bed

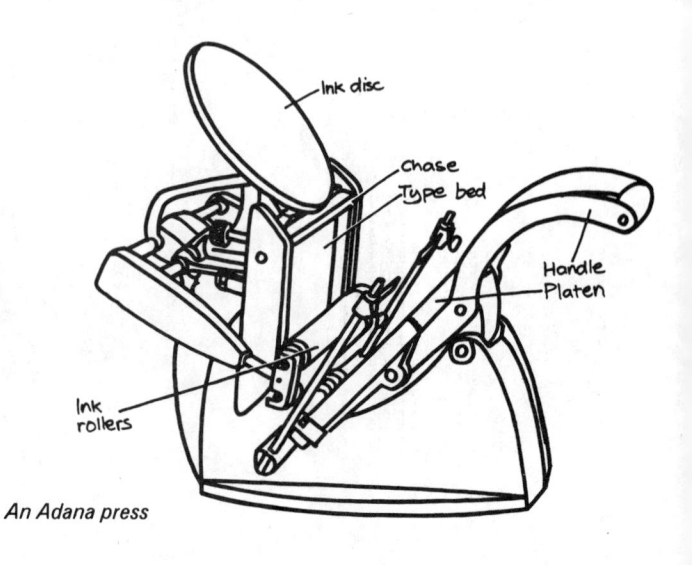

An Adana press

Conference poster produced by letterpress using a wide range of interesting typefaces (61 × 22 cm)

Two-colour cover for poetry book (21 × 14 cm)

- Letterpress is the only process by which you can print on to odd-shaped things (like envelopes) that cannot be fed through an offset-litho press.
- If you need to change part of the information on the job it is probably cheapest to do it by letterpress. For example, if you need tickets with a different date and speaker for each lecture, you can change the date and speaker simply by changing the type, leaving the rest of the information in place.
- Using a letterpress printer may save you time, because you do not have to make a paste-up.

Illustrations. Photographs and line drawings can be used with commercial letterpress, but only by converting them into halftone blocks or line blocks by photo-engraving, which can be costly (⇨ page 97). If you want to print something with photos, consider using offset-litho, which may well be cheaper.

Finding and using a letterpress printer. Some large printers have both letterpress and offset-litho and will be able to advise you about comparative costs. To find a letterpress printer look in the yellow pages (under 'Printers') or in trade directories; or look at other publications like your own and see where they were printed. Some letterpress printers specialize in small things like cards, tickets, invitations and stationery; they are known as jobbing printers. Small local jobbing printers are quite common, particularly in rural areas – in big cities many have been replaced by instant printers. Some stationers act as agents for jobbing printers and are able to take orders for things like printed invitations. They will usually have a sample book to enable you to choose the style and size you want.

82 OTHER PRINTING METHODS

FOR THIS TYPE QUOTE B1
FOR THIS TYPE QUOTE REF B2
For This Type Style Please Quote Reference Number B 3
FOR THIS TYPE STYLE QUOTE REF NO B 4
For This Type Style Please Quote Reference Number B 5
FOR THIS TYPE STYLE PLEASE QUOTE REFERENCE NUMBER B 6
For This Type Style Please Quote Reference Number B 7
FOR THIS TYPE STYLE PLEASE QUOTE REFERENCE NUMBER B 8
FOR THIS TYPE PLEASE QUOTE REFERENCE NUMBER B 9
For This Type Style Please Quote Reference Number B 10
FOR THIS TYPE STYLE QUOTE REF NO B11
FOR THIS TYPE STYLE PLEASE QUOTE REFERENCE NUMBER B 12
For This Type Style Please Quote Reference Number B 13
FOR THIS TYPE STYLE PLEASE QUOTE REFERENCE NUMBER B14
FOR THIS TYPE STYLE PLEASE QUOTE REFERENCE NUMBER B 15

This leaflet shows the styles and sizes of type available from one jobbing printer

The printer will want to see your text before giving you a price for the job, in order to calculate the cost of typesetting as well as of printing. Ask to see some samples of typefaces and sizes, then indicate the arrangement of the text on a rough drawing showing the sizes of text and headings. (⇨ 'Copy-fitting', page 92.)

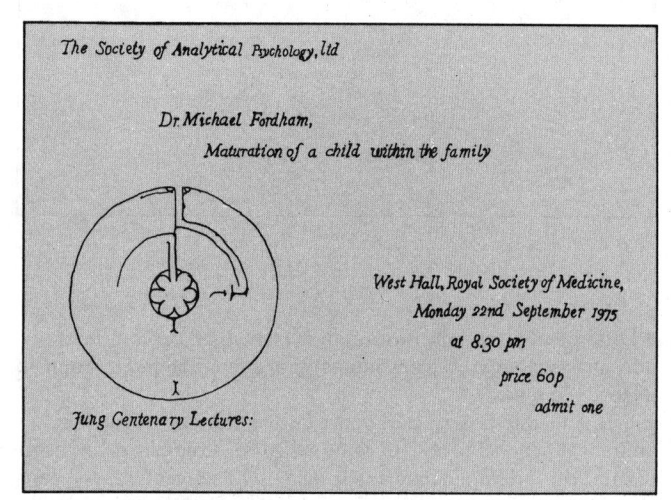

A rough layout to show the printer what is required . . .

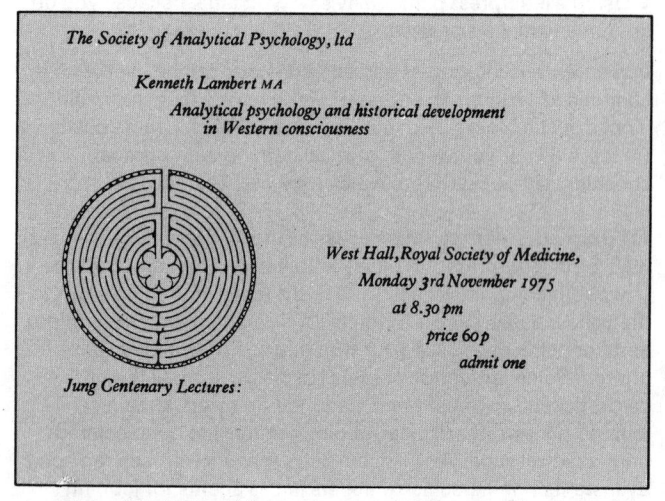

. . . and the ticket printed from this rough (10·5 × 15 cm)

SPIRIT DUPLICATING

The faint purple maps with a smell of methylated spirits that you may remember from school were produced by spirit duplicating (also called 'ditto' duplicating or 'Bandas'). If care is taken this process can give interesting results – it is, for example, the only method by which multicoloured prints can be printed from one master at the same time. It is cheap and easy to do for short runs if the quality is not too important. The image will not be very dark and its edges will be rather blotchy.

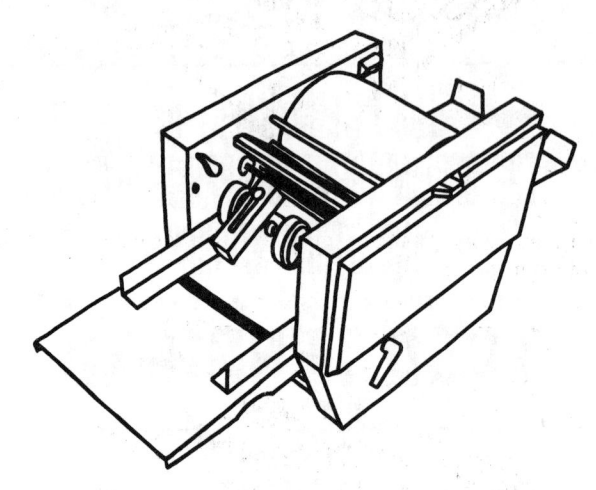

Banda 100 spirit duplicator

HOW IT WORKS
This process uses a dye sheet and a master. The dye sheet is covered with spirit-soluble dye and looks like carbon paper. The master is a sheet of special paper with a shiny side and a matt side. The design is transferred to the shiny side of the master by being typed or drawn on the matt side with a dye sheet lying face up underneath. The pressure of the drawing (or typing) transfers the dye from the dye sheet to the shiny side of the master, thus creating the design in reverse.

The master is then clipped round the drum of the machine. As each sheet of paper enters the machine it is dampened with methylated spirits. As the damp paper passes through under the master it dissolves some of the dye, which is then transferred to the paper to make the print. The paper dries out rapidly after it leaves the machine. Because the dye is being slowly dissolved off the master the prints get fainter and fainter – you can only get 100 to 200 good copies from one master.

COLOUR
The colour of the print depends on the colour of the dye sheet. The colours available are purple, blue, black, red, green, yellow and brown. Some colours print more strongly than others; purple makes the most copies from one master. You can make a multicoloured master by simply changing the dye sheet whenever you want part of the design in another colour. All the different colour dyes on the master are dissolved at the same time as the paper goes through the machine, and all the colours print at once.

Making a multicoloured master

1. Use a red dye sheet to draw in the part of the design you want to print in red.

2. Then change to a blue dye sheet and draw in the rest of the design that is to print in blue.

SPIRIT DUPLICATING

Making the master by hand

1. Lay the dye sheet dye side up on a smooth hard surface with the master sheet matt side up on top. Using a ball-point pen, a pencil or a crayon, draw your design on the top of the master.

2. The image in dye has been transferred in reverse on to the shiny side of the master sheet.

Making the master with a thermocopier. You can make a master from a paste-up by copying it on a thermocopier. The carbon in the paste-up is heated with infra-red and the heat transfers the dye on to the master. Your paste-up therefore needs to be made of materials that contain carbon, such as pencil, typewriter ribbon ink, photocopies or black-and-white printed matter. If your paste-up is made from other (non-carbon) materials, you can make a photocopy of it and use this as the artwork instead.

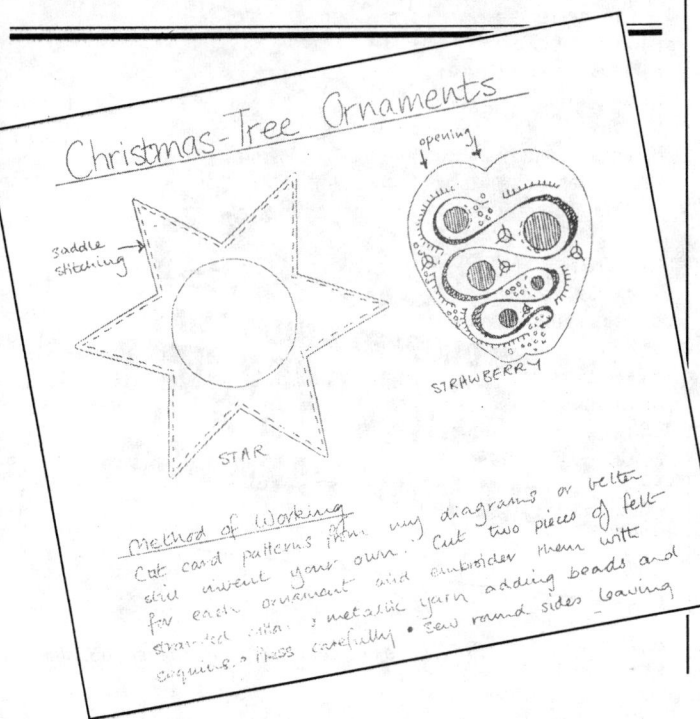

Using a spirit duplicator (e.g., a Banda)

1. Clip the master round the drum dye side up. Stack the paper in the paper tray.

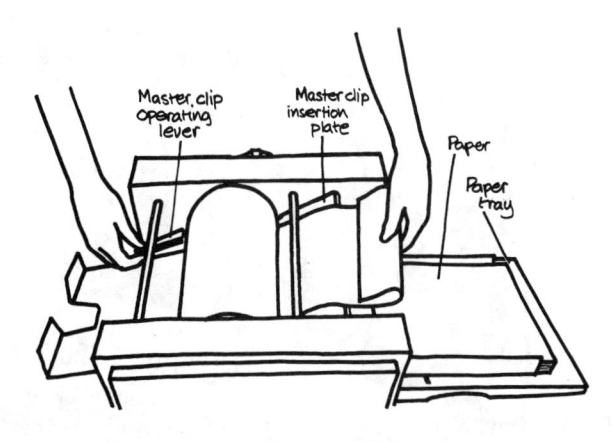

2. Pump up the spirit so that it soaks the damping system. (Not too much to start with, since more will pump up automatically as the paper is fed through.)

3. Switch on the pressure roller. Turn the handle to feed the paper through under the roller and make the print. Adjust the pressure, and if necessary pump up more spirit, until you are getting a good print. (If you make the first prints too dark you will find that you won't get as many good copies, because all the dye will be dissolved off the master too soon.)

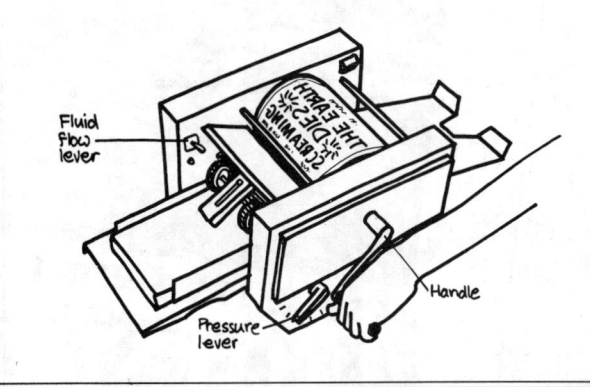

Left: Detail from an A4 leaflet printed on a spirit duplicator

THE ISLINGTON GUTTER PRESS

ISSUE 71
October '80
15p

ISLINGTON'S SOCIALIST MONTHLY

NAG'S HEAD 30 SECS AFTER 1 MEGATON H-BOMB DROPPED 6 MILES AWAY

PROTEST AND SURVIVE!
ALL OUT OCT 26   ASSEMBLE HYDE PARK NOON

Front page of a community newspaper (A3, offset)

Design & Paste-up

Introduction 85
Design 85

Roughs 86

Grids 89

Simple formats 89
Cards 89
Newsletters 89

The text 89
Handwriting 89
Typing 89
Typesetting 90
Copy-fitting 92
Marking up the text for the setter 93

Headlines 94
Ways you can make headlines 94
Photoset headlines 96

Borders and rules 96

Photographs 97
Line images 97
Halftone images 98
Special-effects screens 99
Scaling up photos 100

Drawings, cartoons and other illustrations 100
Line 100
Tone 100

Other types of image 102
Collage 102
Imitations 102

Photo-prints 102

Making the paste-up 103
Equipment and materials 103
Drawing your grid 103
Checking the layout 103
Pasting up 104
Position of photos 105
Paste-ups for more than one colour 105
Tone overlays 105
Reversing out 105

INTRODUCTION

There are two stages in preparing an image for printing. First you must draw out your ideas on paper and make a rough of the design. The design is then either drawn directly on to the plate or stencil, or made into a paste-up which is copied photographically on to the plate or stencil.

This chapter describes the various elements you may need to consider and use in your design; tells you how to produce them; and then shows how to put them all together to make the paste-up. (You may need to consult previous chapters for particular points to bear in mind when making paste-ups for the processes with which they deal.)

DESIGN
Do not be intimidated by designers about what is 'good design' – design your job so that *you* like it, and so that it is readable and makes the message clear. The main things you will need to consider are:

Ideas. What do you want to say? To whom? To express your ideas clearly you may need to give priority to certain parts of the information and you may need headlines to guide the reader, or pictures to show ideas visually or to draw attention to the publication.

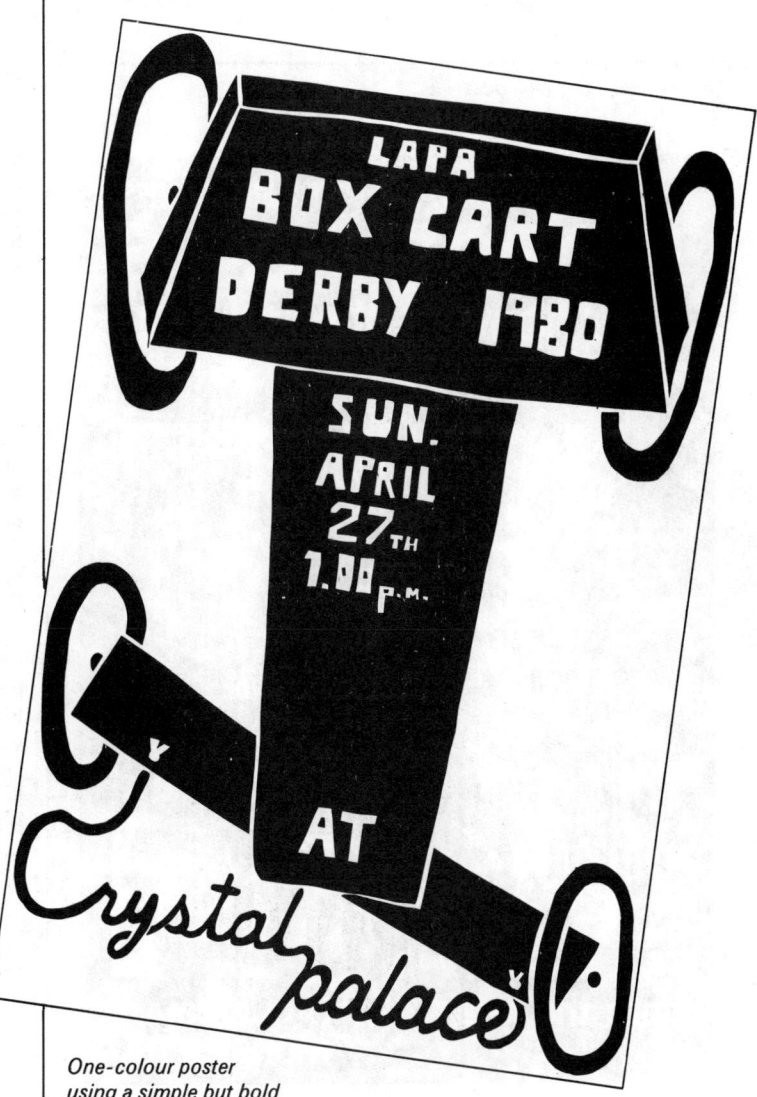

*One-colour poster
using a simple but bold
visual idea (76 × 50 cm, screen printed)*

Newsletter cover (A4, offset)

Magazine cover (A4, offset)

Format. What kind of publication is it – leaflet, poster, pamphlet, magazine, newspaper, book? What size and shape do you want? Will it be sold? Does it need a striking cover? Is it a poster that must be read and understood from a distance? What will catch the eye?

Style. Think about what kind of look you want your publication to have. For example, should it look 'flashy', 'folksy', 'traditional', 'punk-rocky'? What audience do you want to attract? Look around at other publications and see how they present their information.

Print process. Each process has its advantages and limitations, which determine the choice of size and format for your publication and the types of images you can use. This has therefore to be considered before you commit yourself in terms of design and paste-up. (The general Introduction and the introductions to the individual chapters provide guidance to help you choose your printing method.)

ROUGHS

You can decide more or less how you are going to arrange your material by making a small rough drawing, but a full-size rough may be more useful if you want to see how your ideas will look in practice. You may need to make several roughs so that you can try out various ways of fitting in and arranging the information – once you have something down in front of you you can then chop and change it until you find an arrangement that works well.

For simple jobs such as a poster or one-sided leaflet the rough can be made by drawing out the overall areas of text, heading and/or pictures in pencil or felt-tip pen, and where necessary blocking in any large areas of solid or colours. Another way to make a rough is to stick down pictures, areas of text, and headlines etc. cut from other publications; this will help you to see how a page will look.

With more complicated projects like pamphlets or books, you should talk to the printer at this stage about 'imposition' and 'finishing'. Because the printer will be printing more than one page on each sheet of paper, you may have to arrange the pages on the paste-up in a particular pattern so that when the sheets are cut up later the pages come out in the right order – this is called the imposition. (Sometimes you can supply printers with single pages and they will do the imposition.) After printing, the pages have to be folded, trimmed and bound: this is the finishing.

It is also useful to plan the design and artwork by making a mock-up of the actual publication (this is called a dummy), taking into account the imposition and finishing, so that you can see how page will follow page, and how it will all work together. Once you have prepared a rough and worked out the areas that different parts of the design will take up, you then go ahead and make the text and headings, illustrations and so on. At the paste-up stage you can still use this material like jigsaw pieces, moving it around and trying out the pieces in various arrangements. Some people like to work in this way because they can then see the finished thing in front of them, and can make sure that the original design really was the best way to put everything on the page.

Rough for front and back cover of an A4 book about community newspapers

Actual cover for the book printed in two colours (offset)

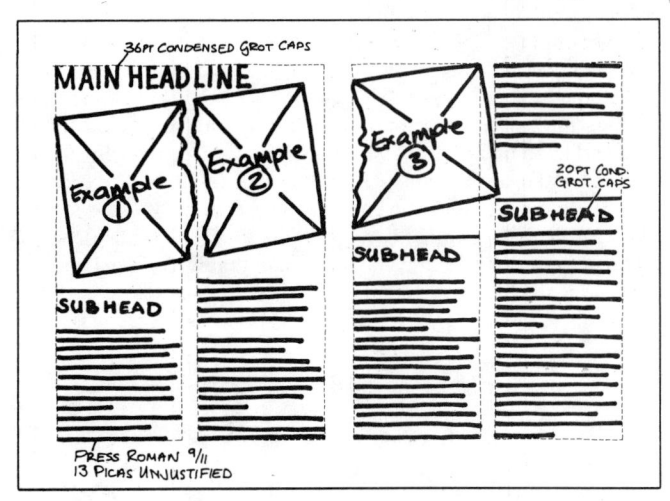

Simple rough for double-page spread of A5 pamphlet

Collaged rough for same double-page spread, using actual headings, text and pictures cut from magazines

Complete Control

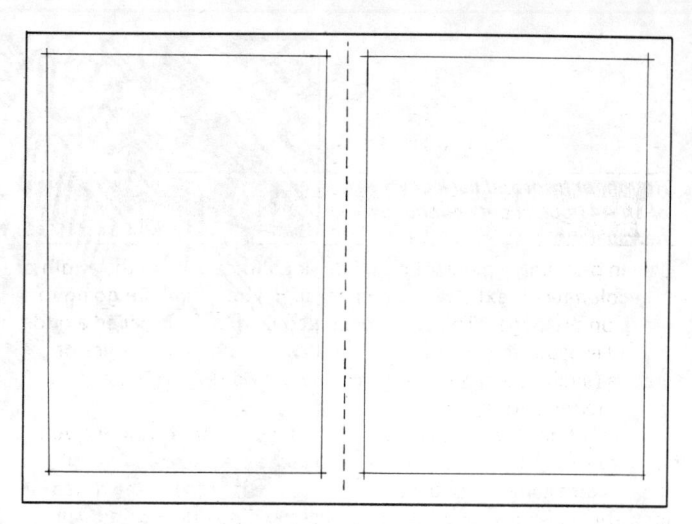

From the moment we're born we become cogs in a relentless brutal machine whose only purpose is to provide wealth and comfort for a small privileged minority — the bosses. Because these bosses control the economy (through their ownership/control of property and of the factories and machines that produce goods) they are able to control every aspect of our lives through science, education, and especially the media (press, radio and television).

We are led to believe that there is no alternative to the present system of things and that our best bet is to tighten our belts and work harder to provide a bigger cake for the bosses in the hope of picking up bigger crumbs for ourselves. That's life! We can either like it or lump it.

Anarchism Says...

Sod this for a laugh! Anarchists believe that the struggle for a better life must involve the destruction of the present system of society and its replacement by a stateless society based on freedom, equality and direct control over our own lives. This means doing away with all forms of domination -- adults over children, men over women, bosses over workers, the young over the elderly and so on.

It also means an end to the wastefulness of production for profit. Children will no longer starve to death in one part of the world while in another large quantities of food are left to rot. In an anarchist society goods will be produced and distributed to the needs of everyone.

Direct Democracy

In the workplace itself methods used and hours worked will be decided collectively by the workers concerned. There will be no bureaucrats sitting behind desks giving orders from on high. Similarly decisions affecting the community -- the planning of schools, recreational facilities, housing, transport etc. -- will be made by everyone in the community.

The whole thing will be co-ordinated not by a central parliament of well-off politicians doing what they want and representing their interests but by workplace and community councils. These will consist of ordinary people elected by their neighbours and fellow-workers.

These people will not be able to make decisions on behalf of everyone else -- their role will be to carry out whatever tasks are delegated to them by the people who elect them. They will have to account for themselves regularly and can be sacked and replaced if people consider they are not doing their job properly.

10   11

Above: Double-page spread of an A5 pamphlet and the grid on which it is based

---

news 6 COMMUNITY ACTION No 49 July-Aug 1980

Wakefield Issues

WAKEFIELD ISSUES No 1 10p

This is the name of a new paper by and for working people in Wakefield. It was initiated by the Trades Council who have two delegates on the collective producing the paper. First two issues have reports on the campaign against discrimination against women by working men's clubs, Snapethorpe hospital campaign, the miners' productivity deal and pay claim, and a campaign against a County Council scheme to re-open a refuse tip which would mean trucks going through a council estate - tenants are asking for trade union support for their planned picketing. 10p plus post from 34 Ashdown Road, Wakefield.

No property boom?

A report from the Property Advisory Group, set up 3 years ago to advise the Department of the Environment on property matters, claims that there is no property boom in sight. It argues that the level of property development, particularly large schemes, will drop and remain depressed for some time. It also examines the role of local authorities in assisting developers to acquire land and comes out against the concept of "planning gain"—councils bargaining for open space, housing land etc in return for permission to build offices and shops. (The Structure and Activity of the Development Industry. HMSO £2.50)

...meanwhile a reunion of asset strippers!

Jim Slater of Slater Walker fame - the fringe bank which went bust in 1975 following the collapse of the property boom, and John Bentley, well known company asset stripper of the early 1970's, have joined forces again. Slater has just bought a £300,000 share in an Irish property company Laganwale Estate of Belfast; and his private property company Strongmead has just made a deal to transform the Tebbitt Group from a tannery concern into a property investment and engineering group. Tebbitt has also acquired two property companies Signgate and Queens Gate. Tebbitts' chairperson is Bentley. So watch out for speculation and more shady deals in your areas

Glasgow Tenants Anti-Dampness Campaign

A meeting of 70 delegates from 25 organisations has decided to set up a Glasgow wide Anti-Dampness Campaign. A follow-up meeting has just elected a committee consisting of representatives from tenants groups in various parts of the city. Aims of the campaign include coordinating action, providing advice and information to individual groups, getting city wide publicity and supporting citywide campaigns and national action. Each tenants group is also to do a survey of their estate to build up a clearer picture of the extent of dampness.

The Glasgow District Labour Party's May election manifesto put dampness top of a list of 12 objectives to improve the quality of council housing - "the persistent problem of dampness in houses will be tackled with vigour and determination to solve this difficult problem". The campaign is demanding additional money from central government to carry out remedial work. Next meeting 23rd August, Secretary- Andrew Kean, 111 Zena St, Provanmill, Glasgow G33.

Construction profits

Despite the drop in new orders and the cuts in public spending the big construction companies are still making big profits. Tarmac, Wimpey, Costain, Laigns, Mowlem, P+O and Newarthill made £200m pre-tax profits last year.

UCATT pamphlet on the cuts

Summarises the cuts in public spending under Labour 1974-9 followed by details of the Tories cuts, particularly those in housing, and their attack on direct works departments. Shows that job security, training and safety standards are all being drastically reduced. Unemployment amongst building workers may double to ½ million. Free from UCATT, 177 Abbeville Road, London SW4

CONSTRUCTION CUTS AND SERVICES

20% rate reduction because of dampness

Tenants on the Mozart Estate in Westminster have just won a 20% reduction in the rate after presenting a case to the Valuation Court about the bad condensation dampness on their estate. The tenants coordinated more than 100 applications for a rate reduction, and , together with the community worker, presented their own case to the Court. In fact, one tenant won a 30% reduction. So now any tenant who reports dampness to the Council will pay lower rates. This victory comes on top of a 15% reduction in the rent which the Association have already won from the council. Contact Hugh Dennis Tel. 01-969 5384

Campaign against school closure succeeds

Parents, residents, teachers and trade unionists in Pill, Newport, South Wales have succeeded in stopping the closure of the Bolt St. School. The County Council voted 41-28 against closure after the campaign had lobbied the education committee, held a packed public meeting at the school, drawn up a large petition and ensured that many local organisations wrote to every member of the education committee detailing arguments against closure. The campaign argued that it was a happy and successful multi-racial school, and that the fall in numbers was only temporary due to redevelopment. They argued that any vacant space should be used to extend much needed nursery education - the campaign is now going on the offensive to get this implemented. Contact Pill-Box, 35 Commercial Road, Newport, Gwent.

Housing Committee, the first since Labour gained control in May. The Cheap Labour Theatre Company entertained the crowd with their spoof of a TV game called the Dampness Game.

Consett fights steel closure

The closure of Consett, Co. Durham steel works is to go ahead making 3700 workers redundant by the end of September. Workers have voted overwhelmingly to fight the closure. The unemployment rate is already 13% and is expected to rise to 30%-40%. An independent audit has shown that the BSC's own projection for Consett includes a £7.5m profit this year. Productivity is nearly twice the BSC average. Detailed report of the campaign in the next issue. Save Consett Steel Campaign, 24 Stockerley Road, Delves Lane, Consett, Co Durham (0207-5 5216 3)

Combined action in Leeds

Three anti-damp groups recently combined to hold a demonstration at Leeds

COMMUNITY ACTION No 49 July-Aug 1980 7 **news**

Crisis City Special

Coventry Crisis Group have produced a 4 page special newspaper challenging the widely held view that the crisis the city is facing is a temporary one. It details the extent of the crisis in Coventry - job losses, rising unemployment, deteriorating housing conditions etc. " The consequences of all this are especially serious because the plant closures and redundancies, the ever growing numbers on the dole, the increased homelessness, rising council house waiting lists, the foreclosed mortgages and mounting rent arrears, the attempt to break trade union organisation and de-skill workers – these are not the result of a temporary recession. What they indicate is a permanent structural change in the economy – the creation of a permanent pool of unemployed labour and a seriously weakened labour movement. They signal a re-organisation of the industrial and service economies so drastic that entire sections of the labour market and, with them, whole communities, will be destroyed. What's taking place, then, is qualitatively different from anything we've seen before, and the consequences for working class cities like Coventry are likely to be devastating."

The Group argue that working class control of living and working conditions is the crunch issue. The idea and the newspaper format could well be used by groups in other cities wanting to inform the labour movement about the root causes of local developments. 5p plus post from Coventry Crisis Group, 40 Binley Road, Coventry CV1.

National campaign against sales

A conference was held in Sheffield in May to start a national campaign against the Government plans to sell council houses at huge discounts. A number of representatives from Labour Groups in Birmingham, Bradford, Liverpool, Leicester, Lambeth supported the Sheffield initiative. However only a few tenants and trade union delegates attended. Various ways of not implementing the new Housing Act were discussed and a committee set up to coordinate further action. Sheffield Tenants Federation, Labour Party and the Trades Council have been holding public meetings and leafletting council estates to explain why council houses must not be sold. Further information from Sheffield Tenants Federation, 69 Division St, Sheffield S1.

Financiers to stop council loans?

A senior member of the Stock Exchange is proposing that financial institutions should stop loaning money to local authorities which the government claim are 'overspending'. Victor Robson, chairperson of Robson Cotterell, a firm of money brokers which invests other peoples' money, mainly from building societies, and takes a fat cut, has threatened to stop loaning £1m a day to these councils.

Action by unemployed

Newly formed Greenwich Action Group on Unemployment has published 'Living on the Dole'. It wants to hear from similar action groups around the country. Contact John Duffield 01 855 2898. Booklet obtainable from G.A.G.U, c/o 105 Plumstead High Street, London SE18 (Tel 01 855 9817). Price 20p plus post.

NEWS SHORTS · NEWS SHORTS · NEWS SHORTS · NEWS SHORTS · NEWS SHORTS · NEWS SHO

Right: Double-page spread of A4 magazine and the grid on which it is based

GRIDS

When planning a piece of print you need to decide on the width of the columns of text and the number of columns you are going to have on the pages. This underlying page structure is called a grid. A grid is especially useful when you have a job with a series of pages (such as a pamphlet, magazine or book) as it gives consistency and regularity.

Instructions for drawing up the grid when pasting up are given on page 103. If you are going to use the same grid a number of times (for example, for a magazine), you can obtain pre-printed grid sheets from your printer or typesetter. The grid is printed in light blue (which will not reproduce if the paste-up is photographed). You paste up your material on these sheets using the grid lines to guide you.

SIMPLE FORMATS

CARDS
Stand-up cards are made from a folded sheet. Position the design correctly so that it will be at the front when the sheet is folded.

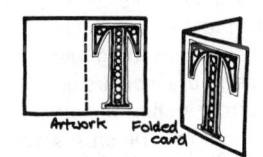

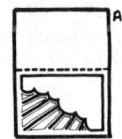

Stand-up cards

NEWSLETTERS
For a four-page newsletter, A5 size (A4 folded):

1. Make a dummy and number the pages.

2. When you open it out flat you will see the correct order for the pages.

3. Paste up the pairs of pages on sheets of A4.

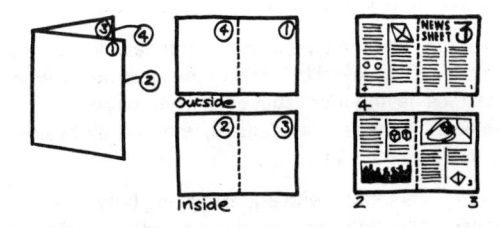

Newsletters or pamphlets of eight pages (or more)
For an eight-page A5 newsletter you will need to make four A4 pieces of artwork:

1. Make a dummy and number the pages.

2. When you separate the pages and open them out you will see the correct position of the pages on A4 sheets.

3. Make the four paste-ups.

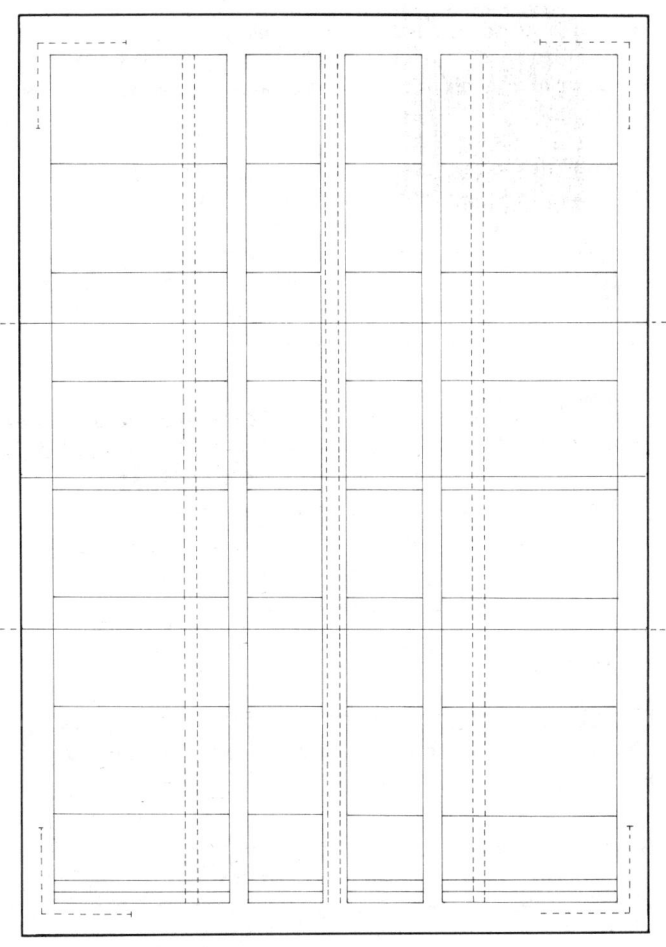

Pre-printed grid, in light blue, from typesetters

THE TEXT

The text for the paste-up can be made in many different ways. It needs to be in a strong black to print well – if it is faint or grey it will not reproduce properly. Do not try to write or type for printing directly on to your paste-up; produce your text material on separate sheets of white paper, to be cut up and then stuck on the final paste-up card or paper.

HANDWRITING
One cheap and easy way to prepare your text is to write it out by hand with a black felt-tip or ball-point pen, with an ordinary pen and ink, or with a special drawing pen. Handwriting is a very versatile method of making text, allowing you to vary the style or the thickness of the writing as you like, and it can give a pleasantly informal look to a publication.

TYPING
Typewritten text takes up more space than typeset text (⇨ page 90) because typewriters have a fixed-width system in which each letter has the same width and takes up the same amount of space

(except for 'executive' typewriters, ⇨ below). Typewritten material can be reduced photographically by up to 20 per cent and still be quite readable. It can also be enlarged for use as headlines (⇨ page 94).

Typing out the text. Use a smooth paper and type out your material in the column width you have decided on. It is usually easiest to type it out in one long column and then cut it up into pieces to fit the pages when you are ready to make the final paste-up (⇨ page 103). Always read through what you have typed; if you have made any mistakes correct them by retyping the whole line, pasting it down on top of the incorrect one.

Manual typewriters

Manual typewriters with old ribbons produce uneven and grey typing which does not reproduce well. If a manual typewriter is the only machine available, clean the keys and use a new ribbon. The blackness can also be increased by laying a new piece of carbon paper face down on top of the paper and typing on to it, hitting the keys hard.

Electric typewriters

Electric typewriters with a carbon ribbon are much better than manual typewriters and usually give an even impression and black print.

Golf-ball typewriters

Golfball typewriters are electric typewriters which can type a variety of different typefaces. Each face is on a separate 'golfball' which can be changed to vary the style. The typing looks better because of the variety of the typefaces, but as these typewriters generally have only a fixed-width system the letters still have a 'typed' look.

'Executive' typewriters

'Executive' typewriters are electric typewriters with a varied-width system, which gives a more readable text because the letters are not all of the same width.

TYPESETTING

Advantages and disadvantages
- Typesetting will make your text look like the printed matter that people are used to reading in magazines and books.
- You can fit a lot more typeset text into the page than typed text because the letters are of varied widths, and the type is also available in smaller sizes than is possible with typewriters.
- The wide choice of typefaces, sizes and weights of type can be used to give emphasis and variety to your text.
- You have to pay for typesetting, but this cost is sometimes covered by the saving you can make on the printing costs, as typeset text will fit into fewer pages.

Typesetting terms and measurements. It is helpful to understand the following terms when talking to your printer or typesetter.

Type measurements. The *point* is the standard unit used for measuring both size of type and the size of the spaces between lines of type. There are approximately seventy-two points to the inch. The *pica* is the unit used for measuring the length of lines of type. There are twelve points in a pica; and therefore six picas to the inch.

Type size. There is a very large range of type sizes. You usually make your choice according to the kind of information the typeface is being used for. A very rough guide is:

6–8 pt	For minor information, captions etc.
8–12 pt	For the main text.
10–18 pt	For introductory paragraphs, headings and subheadings in text etc.
30 pt upwards	For large display headlines for posters, leaflets, covers etc.

Leading. The space between the lines of type is called the leading (pronounced 'ledding'), and is measured in points. The leading may be varied according to the amount of space that is available for the text, and so that it is easy to read the text when it is set on a given length of line. If you need to squeeze the text into a small space you may have to set it without any leading – it is then said to be set solid. Ten-point type set with 1-pt leading is written down as 10/11 pt (and said to be 'ten on eleven point') and 10-pt type set with 2-pt leading is written 10/12 pt, and so on.

The column width. The length of the line of type in a column is called the 'measure' or column width, and is calculated in picas. To measure it, you may need to get a typesetting ruler (known as a 'type-scale' or 'printer's rule') but if you are unable to the typesetter will be able to convert from inches or metric measurements. Note that ideally there should not be on average more than ten to fifteen words to the line: lines with more words than this are harder to read.

Typefaces and weights. Typesetting is available in a large variety of sizes and styles (called typefaces). Most typefaces are designed in at least bold, medium, light and italic forms.
Each typeface has a name, such as Univers, Times Roman or Baskerville. You can give your text a unity and consistency by using the same typeface in various sizes and forms. Bold and italic are often used to give particular emphasis within text set in a light or medium typeface.
Other typesetting terms include the following:

Justified setting. When the type is set so that each line is of the same length (this is achieved by varying the spacing between words, some lines having more space between words than others). The result is the familiar column, with a straight side on the right as well as on the left.

Unjustified setting is setting with equal spacing between the words, as a result of which each line has a slightly different length. This produces an irregular right-hand side. It is also called *ranged left* or *flush left* (*ragged* in the U.S.A.). This book has been set unjustified, for example. The text can also be set *ranged right* (with the irregular side on the left). *Centred setting* is setting in which each line is centred on the setting area.

Typesetting systems. There are many different systems, but the ones listed below are the most useful.

IBM composing machines (e.g. Selectric, Electronic) are the simplest and most common small machines now available. They are often used by individuals, small companies, cooperatives and community printers. Some of these machines cannot justify the text (⇨ above) automatically, which means that you would have to pay more for this.

Varitypers are older machines, similar to IBM typesetters. They produce lower-quality setting but have a larger range of typefaces.

Two families of type

Univers Light No. 45
ABCDEFGHIJKLMNOPQRSTUVWXYZ
abcdefghijklmnopqrstuvwxyz
£1234567890&.,:;'"‐"!?()*

Univers Light Condensed No. 47
ABCDEFGHIJKLMNOPQRSTUVWXYZ
abcdefghijklmnopqrstuvwxyz
£1234567890&.,:;'"‐"!?()*

Univers Medium No. 55
ABCDEFGHIJKLMNOPQRSTUVWXYZ
abcdefghijklmnopqrstuvwxyz
£1234567890&.,:;!?""""‐—()*⁄

Univers Medium Italic No. 56
*ABCDEFGHIJKLMNOPQRSTUVWXYZ
abcdefghijklmnopqrstuvwxyz
£1234567890&.,:;!?""‐—()*⁄*

Univers Bold No. 65
**ABCDEFGHIJKLMNOPQRSTUVWXYZ
abcdefghijklmnopqrstuvwxyz
£1234567890&.,:;'""""!?‐—()*⁄⁄**

Univers Extra Bold No. 75
**ABCDEFGHIJKLMNOPQRSTUVWXYZ
abcdefghijklmnopqrstuvwxyz
£1234567890&.,:;'"‐"!?()***

Plantin Light
ABCDEFGHIJKLMNOPQRSTUVWXYZ
abcdefghijklmnopqrstuvwxyzffﬁﬂﬃﬄ
£1234567890&.,:;'"‐"!?()*

Plantin Light Italic
*ABCDEFGHIJKLMNOPQRSTUVWXYZ
abcdefghijklmnopqrstuvwxyzffﬁﬂﬃﬄ
£1234567890&.,:;'"‐"!?()**

Plantin Bold
**ABCDEFGHIJKLMNOPQRSTUVWXYZ
abcdefghijklmnopqrstuvwxyzffﬁﬂﬃﬄ
£1234567890&.,:;'"‐"!?()**

Plantin Bold Italic
***ABCDEFGHIJKLMNOPQRSTUVWXYZ
abcdefghijklmnopqrstuvwxyz
£1234567890&.,:;'‐!?()***

Plantin Bold Condensed
**ABCDEFGHIJKLMNOPQRSTUVWXYZ
abcdefghijklmnopqrstuvwxyzffﬁﬂﬃﬄ
£1234567890&.,:;'"‐"!?()***

Photosetting. This is a complex system which uses a photographic method to set the type. At the moment photosetting is often more expensive than other systems, but this is a rapidly developing field and prices may come down as the technology develops. Photosetting gives high-quality setting in a very large range of typefaces and sizes. The type size can be varied easily, which makes it possible to set subheadings, for example, in a larger size in the correct position in the text. (This saves pasting in a lot of small headings.)

Finding a typesetter. To find a good typesetter ask around, look at other publications to see who typeset them, or look in the yellow pages. For small jobs almost any setting will do if it is available locally and not too astronomical in price, but for larger projects one particular typesetting system or company may be more suitable than another. Projects with large amounts of text, such as books, are often set in metal type and printed by letterpress (⇨ page 79). Before you go ahead with a typesetter make sure that you know how much it is going to cost. For straightforward text the price will depend on the typesetter and the type of job. 'Community' typesetters in London charge between £6 and £12 per 1,000 words, while larger commercial firms charge from £15 upwards. In the U.S.A. typesetting costs from $8 to $15 per page, depending on the number of characters per page. (Some typesetters quote a price per 1,000 'keystrokes' or 'ens', which means 1,000 characters (or letters) plus the spaces between them.) There will be extra charges (or hourly charges) for lists, tables or badly prepared copy, and some companies also have a minimum charge for small amounts of setting.

Redesign
PHOTOSETTING

The following is a list of typefaces available on our ITEK photosetting system. The maximum column width set is 48 ems (picas).

Helvetica Light
Helvetica Medium
Helvetica Italic
Helvetica SemiBold
Helvetica SemiBold Italic
Helvetica Extra Bold
Avant Garde Bold
Avant Garde Bold Condensed

Century Medium
Century Italic,
Century SemiBold
Century SemiBold Italic
Times Medium
Times Italic
Times SemiBold
Times SemiBold Italic

All typefaces are available in the following point sizes:
5¾, 6, 6½, 7, 7½, 8, 8½, 9, 9½, 10, 10½, 11½, 12, 13, 14, 15, 16, 17, 18, 19, 20, 21, 22, 23, 24, 25, 26, 27, 28, 29, 30, 31, 32, 33, 34, 35, 36.

9 London Lane, London E8. ☎01-533 2631

Typesetter's leaflet showing the styles and sizes of type available (A5)

It is a good idea early on to talk to the typesetters about the typefaces, weights and sizes they have available and to discuss any particular problem you think may arise from your text. They will usually be able to give you samples to look at and these will also help you to calculate roughly what size type you should use to fit your text into the space available. For more accurate calculations you will need to do some 'copy-fitting' (see below).

COPY-FITTING

If you only have a small amount of text to be set and plenty of space to fit it in, then you could go straight ahead and get it typeset in the size and measure you prefer. But when you have larger quantities of text and/or limited space you need to calculate more exactly the space that your text will take up once it has been set in a particular typeface and size. Copy-fitting is the process of working out how to fit the typesetting into a space in the design. It is also known as casting off – especially when you have a text and are working out how much space it will take.

The following is one method of casting off:

1. *Find out the total number of characters* (letters) in the text by multiplying the average number of characters in the line by the number of lines of text. Note that the spaces between the words and the punctuation marks count as characters. For long texts this process can be done more approximately by multiplying the average number of characters on a single page by the total number of pages of text. (This is easier to do if you are working from typewritten text.)

2. *Decide on the measure* you are going to use – this is quite easy because it is usually dictated by the format of the page. For example, an A5 page could take one column 4½ in. wide (27 picas); an A4 page could take two columns 3 in. wide (18 picas).

3. Next *decide on a typeface and size.* The manufacturers of typesetting machines publish 'copy-fitting tables': using these look up the number of characters that will fit into your measure.

4. *Divide the total number of characters* in the text *by the number of characters per line.* This gives *the total number of lines of text.*

5. *Calculate how much space* this number of lines of text will take up. An easy way to do this is to look at something that has already been set in the same-size typeface and leading (probably in the samples that the typesetter gave you). Or you can use what the professionals use – a type-scale or Haberule; this looks very complicated but is very easy to use. Lines are marked on the scale to show how much space lines of setting take up; sizes from 6 pt to 14 pt are shown. By placing the scale alongside the space available you can read off the number of lines of text that will fit in varying sizes and leading.

The type-scale also helps you calculate the space when you are using leading (extra space) between the lines. If you are using 1-pt leading, add this on to the size of type; for example, if you are using 10-pt type set with 1-pt leading you use the 11-pt column on the scale.

If you have too many lines of text to fit into the space do the copy-fitting again with:
- a smaller-size typeface, *or*
- less leading, *or*
- a wider column width, *or*
- edit your copy.

If your text does not fill the space try using the extra space for larger pictures, headlines, etc., or leave space, or do the copy-fitting again with:
- a larger-size type, *or*
- more leading, *or*
- a smaller column width.

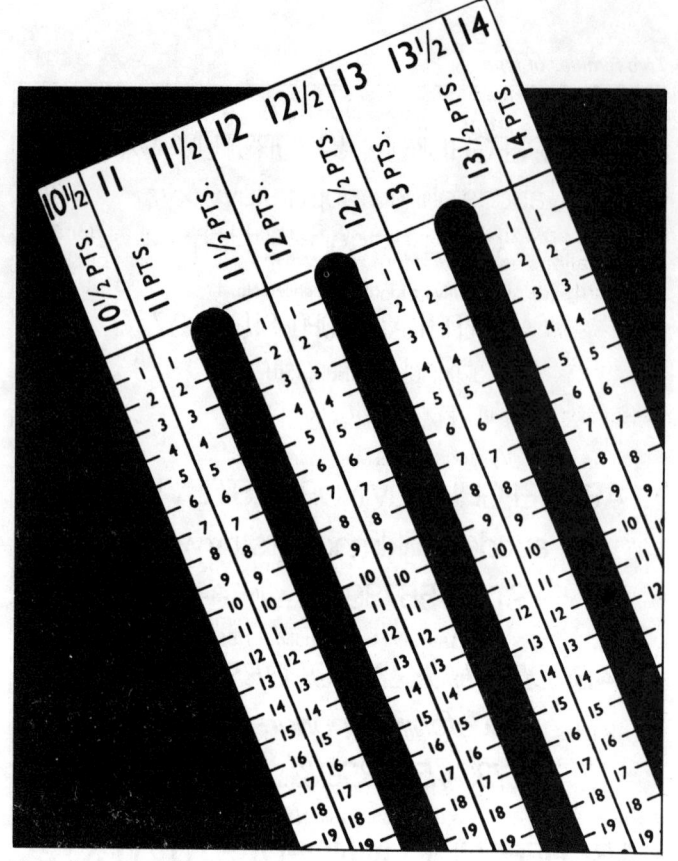

Part of a type-scale

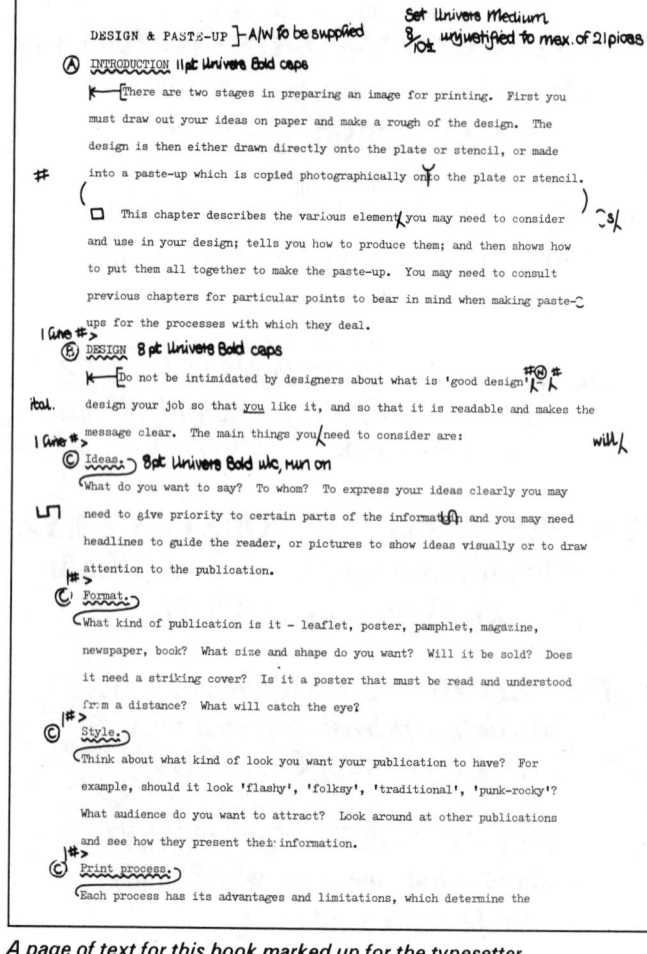

A page of text for this book marked up for the typesetter

If your type size, leading and measure are already decided on (as in the case of a magazine) then you will have to consider cutting some text if there is too much, or filling up extra space with more text, illustrations or the like.

Simpler copy-fitting. Less accurate copy-fitting can be done by calculations using the average number either of words per line or of words (or characters) per column inch:

Words per line. Work out the average number of words per line in a sample of the chosen type. Then divide the total number of words in your text by this number to find the number of lines of typeset text you will have.

Words per column inch. Working out the number of lines of text in terms of column inches is quite common and saves time, especially if you use a particular typeface, size and measure regularly (as in a magazine). Make an average word (or character) count of an inch of column in some existing typesetting that matches what you are going to use. This gives an approximate number of words (or characters) per column inch. Divide this into the total number of words (or characters) to find out how much text you will have in terms of column inches.

MARKING UP THE TEXT FOR THE SETTER
Type out your text so that it is clear and easy to follow. Type on one side of the paper only and number the pages. Leave wide side margins, and use double spacing between the lines. It is a good idea to make a copy of the text to keep for yourself just in case the typesetter's copy gets lost.

You must indicate on the text what you want the typesetters to do. Professionals use a lot of special symbols to do this (⇨ table below), but if you write clearly what you want on the top of the text that should be adequate. If possible talk to the typesetters as well, to make sure that they understand your instructions.

Note that the typesetters will correct at no extra charge any mistakes they made in setting your text. However, they will charge extra for any changes you make in your proofs that are not just corrections of their mistakes. Such changes can therefore prove expensive, so try to supply, if possible, copy that has been agreed upon by all concerned as final.

On the first page specify all the things that will be constant throughout the text:
- typefaces and sizes
- leading
- the measure (the width of the column)
- justified/unjustified
- any extra spacing between paragraphs
- indenting of paragraphs

and write: 'Follow throughout (unless otherwise specified), please'

Other information is written in coloured pen or pencil at the relevant point in the text. Anything, such as these instructions, that is not to be printed should be circled, to leave no doubt that it is not to be set. In correcting proofs, use red for printer's errors and blue for any changes you make that depart from the original text. The table shows a few symbols that you may need to use.

Symbols for marking up text or correcting proofs

Mark	Meaning of mark	What to put in the margin*	What to put in the line
⌣⌣	Set in italics	⌣⌣ or *ital.*	The relevant characters are underlined
∿∿	Set in bold	∿∿ or **bold**	Written under the relevant characters
≡	Set in capitals	≡ or caps	Written under the relevant characters
⌐ or ⌈	New paragraph	⌐ or n.p.	Inserted between the relevant sentences
Y or #	Insert space	Y or #	/ between characters
⌒	Close up	⌒	⌒ between characters to be closed up
—(or >	Add space between lines	—(or >	Extend between the lines
—⟩ or (	Less space between lines	—⟩ or less #	Extend between the lines
⌐	Move type to the right	⌐	⌐Enclosing type⊢→
⌐	Move type to the left	⌐	←⊢Enclosing type⌐
⊔⊓	Transpose matter shown	⊔⊓	⊔⊓ Showing matter to be transposed
⌒	Run on (no new paragraph)	⌒ or run on	Links the two paragraphs
⌿	Delete character or words indicated	⌿	/ or ⊢—⊣ through the character or characters
⋏	Insert new matter	⋏ (new matter)	⋏ (new matter)
≢	Change to lower-case letters	≢ or l.c.	Circle relevant characters

* The symbols need only be used in the margin when correcting proofs.

HEADLINES

In text. The reader often needs to be guided through a long piece of text by headings and subheadings which break the information into separate sections. Headlines are also useful in giving character to your publication and visual interest to the text.

In posters and leaflets. Headlines are used in a slightly different way in posters and some kinds of leaflets with little or no text – which are virtually all headlines. To convey the most important information decide on the priorities and use your large or bold lettering for this and smaller lettering for the remainder.

Whichever way you make your headlines do them on separate sheets and cut them out and stick them on the paste-up.

WAYS YOU CAN MAKE HEADLINES

Hand-drawn headlines. One cheap and simple way to make headlines is to draw them by hand. You can design your own letters, or trace them out of books (available at libraries). Rough out the letters first in light-blue pencil (which will not reproduce when the paste-up is photographed) to see whether they will fit into the space available. Then draw them out in black ball-point or felt-tip pen, on a piece of white paper. (Use tracing paper or layout paper if you are making a tracing.)

by hand

tracing

Cut-out letters. You can paste down cut-out letters or words (black) from newspapers or magazines, although it may be hard to find all the words you want in the same size.

cut Out letters

Stencilled headlines. Many different styles of stencils are sold at art shops and stationers. They are made of plastic, card or metal. Some are made for use with special drawing pens (e.g. Rapidographs), but metal ones are big and chunky, for use with a large brush or felt-tip pen. Always use black ink or paint.

STENCILS

Dry-transfer lettering. This is probably the most accessible way of making professional-looking headlines. Various brands are available (e.g. Letraset, Mecanorma, Chartpak) from art shops and stationers. Ask to see the catalogue and choose the style and size you want (there is a large range). They are very expensive (£4/$5 a sheet), so make sure you choose wisely!

Dry Transfer Lettering

Using dry-transfer lettering

1. On a sheet of good-quality white paper, rule a guide-line in light-blue pencil and tape the paper down to the table or drawing-board with masking tape. Position the first letter so that the guide-line on the sheet of lettering is lined up with the guide-line you have drawn.

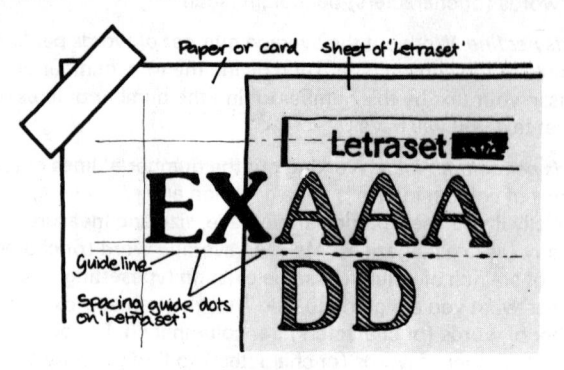

2. Transfer the letter on to the paper by rubbing it evenly with an old ball-point or a soft pencil until all the letter has gone grey, which means that it is stuck down on to the paper.

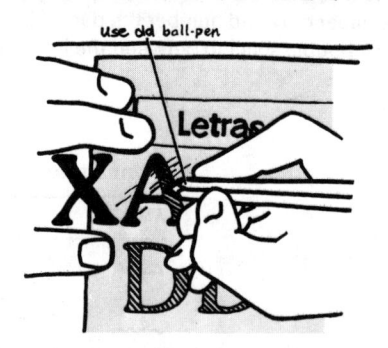

3. Continue until the word is all done, then burnish the lettering by laying the backing sheet on top of it and rubbing it firmly with the rounded end of the art-knife or pen.

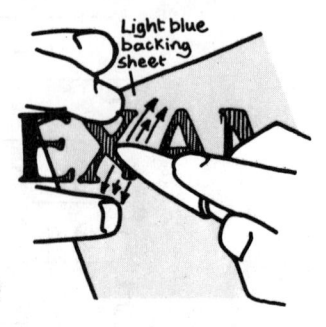

4. Mistakes can be removed by lifting the letters from the paper with a piece of masking tape or with a Cow gum rubber or rubber-cement pick-up.

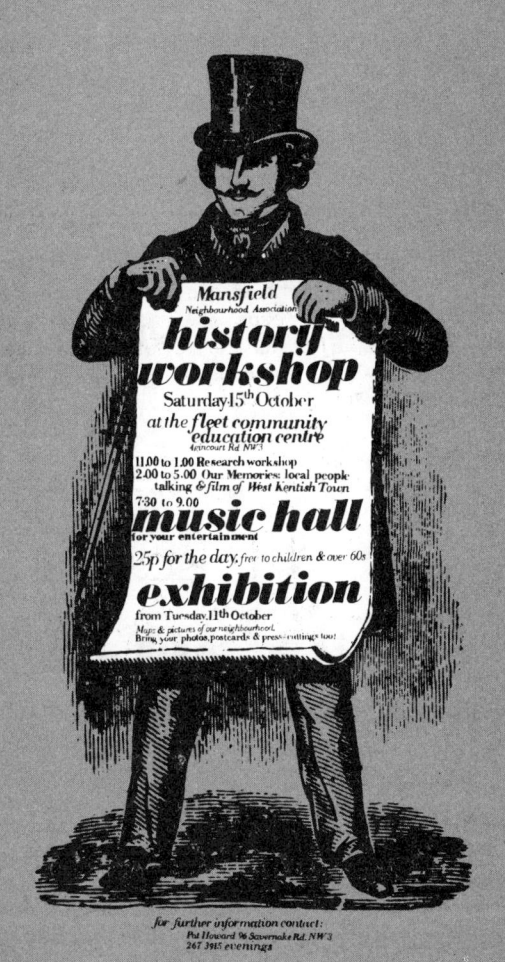

Poster using traced lettering (A3, offset)

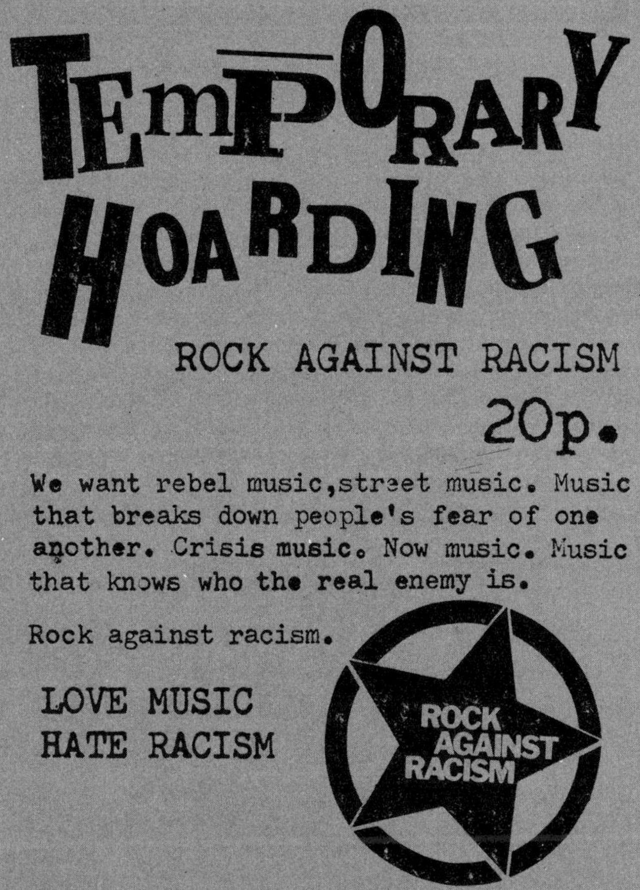

Cut-out letters used on a magazine cover (A4, offset)

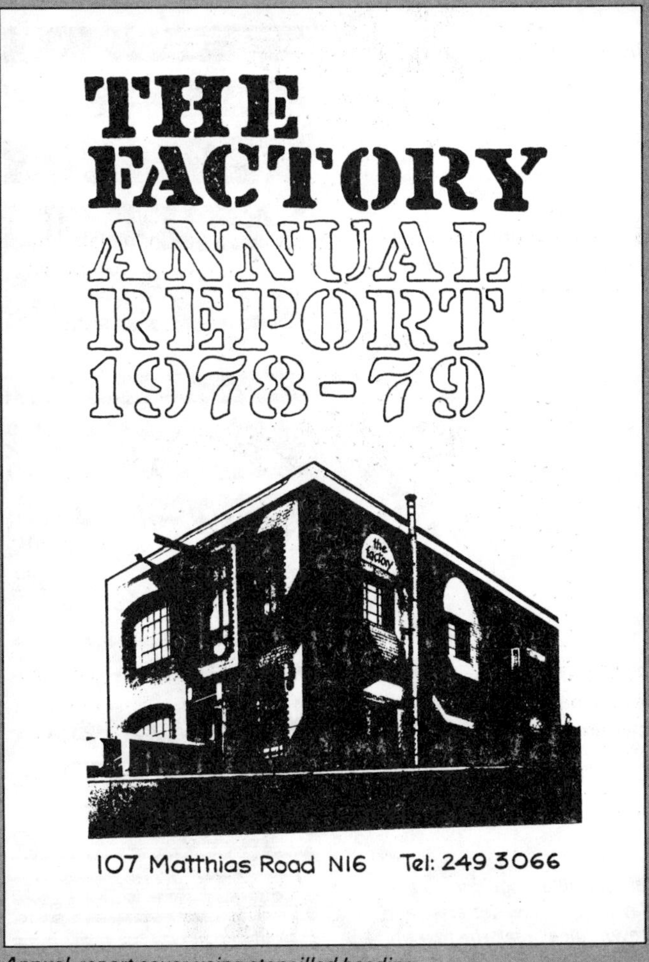

Annual-report cover using stencilled heading (A4, stencil duplicating, two colours)

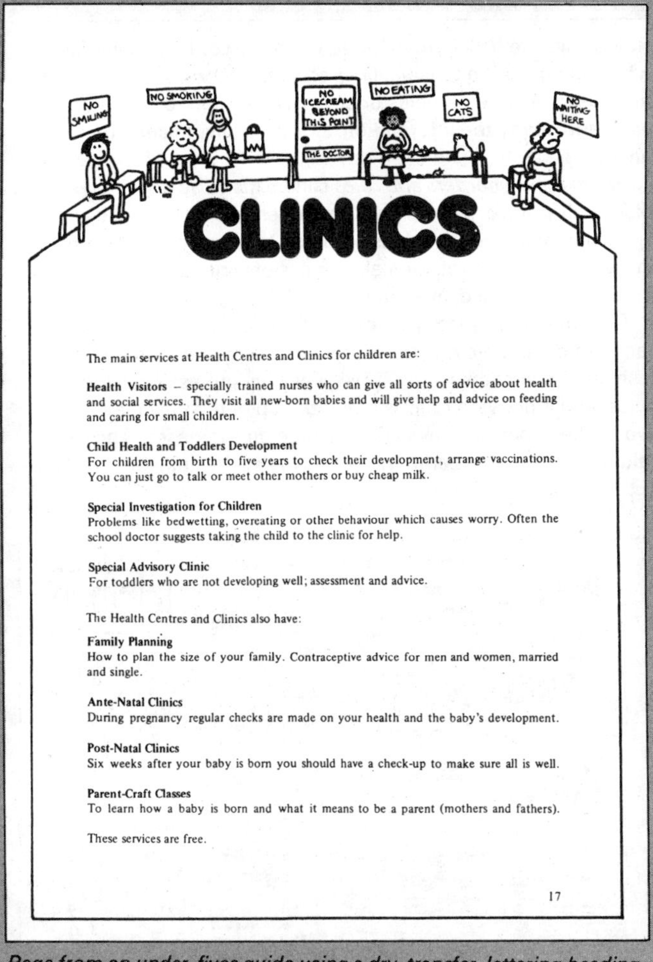

Page from an under-fives guide using a dry-transfer-lettering heading (A4, offset)

96 DESIGN AND PASTE-UP

PHOTOSET HEADLINES

Photosetting usually costs more than dry-transfer lettering, but it may save time and it enables you to have headlines set, by the typesetter, to exactly the size you want. If the text is to be photoset, then you might be able to have the headlines set at the same time, in the correct position. This saves having to paste in all the headlines afterwards. Check with the typesetter that this is possible (it will depend on the type of machine that they are using). You can order your type from the typesetter in the usual way (by style, weight and size, ⇨ page 93), but an advantage of photoset headlines is that you can also order to a cap height (in mm) or to an overall line length – many photosetters in fact work this way rather than to point sizes.

Photosetting

Right: Photoset text and headings on a publicity leaflet for a magazine (A5, offset)

BORDERS AND RULES

Ruled lines are called 'rules' so as not to be confused with 'lines', which in typesetting means lines of text. Decorative borders and rules used between articles or columns can help to divide up the page and clarify the information, as well as giving character to the page.

Hand-drawn borders and rules can be made by drawing with a black ball-point or felt-tip pen. For better-quality rules use a special drawing pen (such as a Rotring); these pens, which come in a variety of nib sizes for making rules of various thicknesses, produce a very solid, even rule.

The companies that make dry-transfer lettering also make a range of decorative *dry-transfer borders and rules*, which come on a sheet or sometimes a roll. This gives very professional-looking rules, and if drawing rules is not your cup of tea they may be worth the expense; they also save time on a large job. They are laid down on the paper in the same way as dry-transfer lettering (⇨ page 95).

Right: Some Letraset borders

Left: Double-page spread for a playgroup's newsletter – hand-drawn rules break the text up well and hand-drawn headings add interest (A4, stencil duplicating)

PHOTOGRAPHS

Most printing methods are unable directly to reproduce the shades of grey in ordinary photos (called 'continuous tone'); photos have to be either converted into 'line' or 'screened' into 'halftone' dots, to make them suitable for reproduction.

LINE IMAGES
The line process picks up only black or white, so that the greys in the photo become either black or white. Photos reproduced in this way become very bold and graphic (although much of the detail is lost). This can be very useful if you need bold images for posters, covers, etc. Use photos which already have strong contrast, to get the best results.

There are three ways of converting photos into line:

1. Continuous-tone photos can be laid straight down on the paste-up and converted into line images when the paste-up is photographed to make the plate (or stencil). This is taking a bit of a gamble because you will not know what the photo will look like until it is on the plate, but it may save you time and the expense of using one of the other methods. Discuss this with the printer beforehand (some printers will not do it).

2. Get a line photo-print made on a process camera from your continuous-tone photo (⇨ page 102), and then paste this down

Annual-report cover using line photos (A4, screen printed, two colours)

on the paste-up. The photo can be enlarged or reduced at the same time, if necessary.

3. If you have access to a dark-room and a basic knowledge of photography, you can convert the photo yourself by exposing your negative on to 'line' paper or film (e.g. Kodak Kodalith), or on to hard photographic paper (grade 4 or 5). The paper print is then pasted down on the paste-up (or the film image used for the positive, in the case of screen printing).

HALFTONE IMAGES

To reproduce the full range of grey tones and all the detail of your original photo you need to covert it to 'halftone', in which the greys are formed by a pattern of black and white dots. This process is done on a process camera and is called 'screening', as it is done by exposing the photo through a dot screen.

Different-sized dots (defined as number of dots per inch) are used depending on the quality of reproduction you need – the smaller the dots, the more detail is kept. Different-sized dots are required for different printing processes:

Screen printing: 20–65.
Simple offset-litho (instant printing) or when printing on
    newsprint paper: 65–85.
Better-quality offset-litho: 85–120.
High-quality offset-litho (on coated paper): 120–150.

The photograph below has been printed in four separate parts to show the range of dot sizes that might be used, depending on the printing method employed and the quality required (40, 65, 85 and 120 respectively, reading from left to right)

You can get your photo screened in the following ways:

1. You can rephotograph a photo on a process camera, to make a screened photo-print. This can be put in place on the paste-up and a plate can easily be made for the whole of it in one go.

2 If you need better-quality reproduction the photo and the rest of the paste-up may each be photographed separately by the printer at the plate-making stage. The photo can be enlarged or reduced at this stage if necessary. A screened negative is then made of the photo and a line negative of the rest of the paste-up. The two negatives are then taped back together – this is called 'stripping in'. In this case, when you give the job to the printer all you need to do is to indicate on the paste-up where the photo is to go (⇨ page 105).

3. You can use black-and-white photos that have already been screened and printed in magazines, newspapers etc. This is a convenient source of pictures, but the quality will be unreliable. Stick the photos down on the paste-up, taking care not to get glue on them since they are likely to smudge when you try to remove the glue! (You ought to get permission to use such pictures from the copyright-holder.)

4. If you have access to a dark-room and a basic knowledge of photography, you can make your own screened photos from the negative. The prints you make will not be of high quality but they will be much cheaper and quite adequate for many things (and particularly useful for making film positives for screen printing):

● Using an enlarger, expose your negative through a commercially made 'contact screen' (e.g. Kodak PMT grey screen) on to line film or paper (e.g. Kodalith). See the Kodak manuals for more detailed instructions.

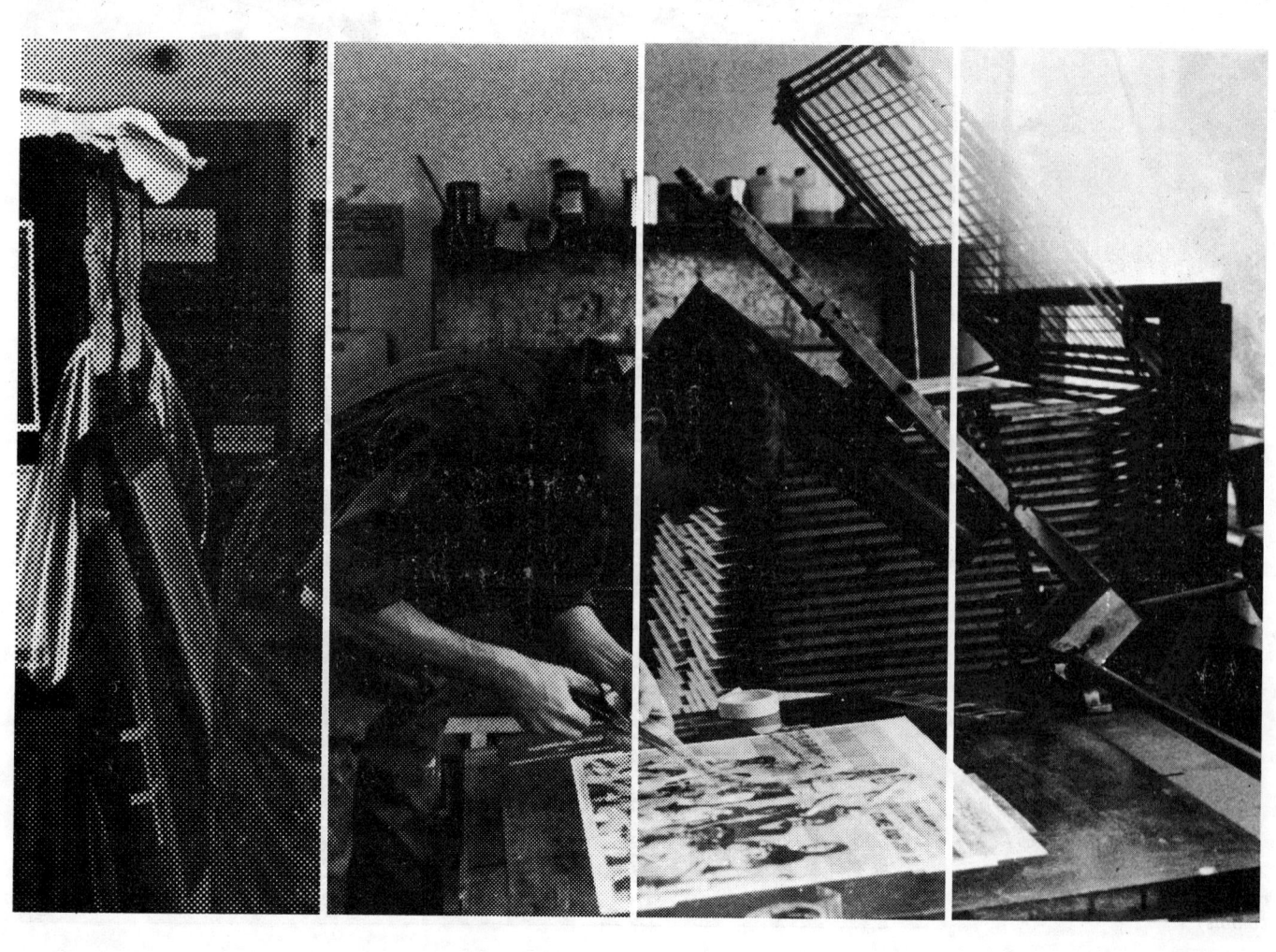

Already screened photos used in a collage illustration (A4, offset)

● A similar effect can be produced by using a sheet of 'instant tone' (made by the dry-transfer-lettering companies) as a contact screen. Choose the one that will give the size of dot your require. Because all the dots will be of the same size it will not look as good as real screening, in which the dots are all of different sizes.

● You can expose your negative on to a film that has a dot structure built into it (e.g. Kodak Autoscreen). Make a contact print of your negative on to the film – this will produce a positive with 122 dots to the inch. To get a larger-size dot, reverse the screened positive back to a negative by contact printing and then enlarge this negative to the finished size of print you want (on film or paper).

SPECIAL-EFFECTS SCREENS
Photos can also be screened into patterns other than dots. Some companies that make photo-prints offer this service; alternatively, you can do this yourself in the dark-room using instant tone (⇨ page 100).

A sheet of 'instant tone' used as a crude dot screen (more easily seen in the enlarged inset detail of sleeve)

Special screen used on the front of a community newspaper (32 × 22 cm, offset)

SCALING UP PHOTOS

When you are getting a photo screened by the printer and enlarged (or reduced), it is useful to know what height it will be when enlarged to the width you have specified. You can then leave the exact space for it in the paste-up.

One way to calculate this is by using a diagonal line:

1. Trace the shape of your photo and then draw a diagonal line from the bottom right corner (A) to the top left corner (B). Extend this line if the photo is going to be enlarged.

2. Measure off (A to C) the dimension to which the photo is to be enlarged.

3. Then draw a vertical line upward from this point (C) until it crosses the diagonal (D) . . . the measurement from C to D is the height of the photo when it is enlarged to the width you specified.

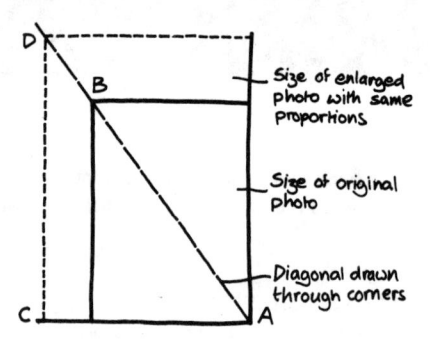

Scaling up a photo

DRAWINGS, CARTOONS AND OTHER ILLUSTRATIONS

Almost anyone can produce a cartoon or drawing to make text more interesting and to help convey information. Since an illustration can often give information better than text, drawings should not always be thought of as something tacked on as decoration to the text; drawings themselves can tell a story – think how popular cartoon strips are! Photographic methods of making plates and stencils mean that most illustrations can be easily reproduced in line or halftone, as we have just seen.

LINE
All illustrations except those containing greys reproduce easily along with the text. Use black felt-tip or special drawing pen for your drawings, and paste them in position on the paste-up with the rest of the information.

TONE
Drawings which contain shades of grey (for example, pen-and-wash drawings or pencil sketches) cannot be reproduced as line because much of the subtle detail would be lost when the greys were converted into black and white shapes. To keep the grey tones and details, the drawing has to be converted into halftone dots in the same way as for continuous-tone photos (⇨ page 97).

Tones and textures. By making patterns of dots, lines or textures, you can add areas of tone to illustrations.

By hand. You can do this by hand by drawing in a pattern of lines or other textures in areas where you want it. This can be useful if you want tone but cannot print a solid (for example, in stencil duplicating if you want to print large lettering, ⇨ page 16).

Instant tone. Companies that manufacture dry-transfer lettering also make sheets of tone (e.g. Letratone); there is a wide range available, from straightforward dots to patterns and cross-hatching effects – consult the catalogues. The tones come in a variety of greys, expressed as a percentage of black (which is 100 per cent).

How to use instant tone

1. Without cutting through the backing sheet, cut out with an art knife an area of film slightly larger than is needed.

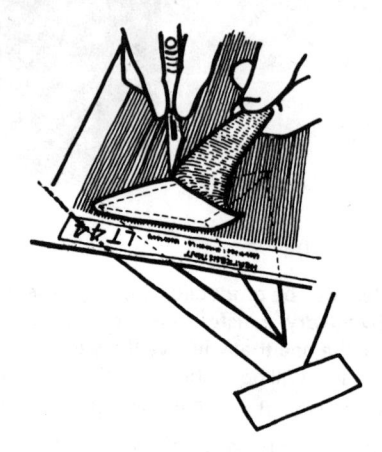

2. Peel the film from the backing sheet, place it in position over the shape and smooth it down lightly.

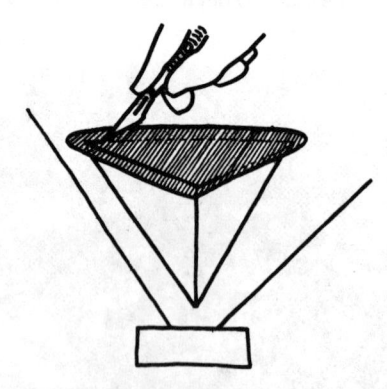

3. Cut around the required shape with the knife and peel off the excess film, then burnish the film down firmly.

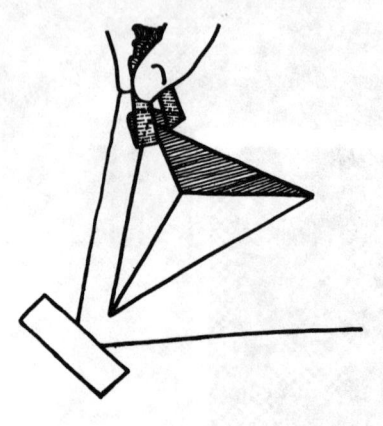

NB Make sure you do not get specks of dirt under the tone as these will be picked up and reproduced on the print.

Cartoon strip used on leaflet (A4, stencil duplicating)

Right: Simple drawings and hand-drawn lettering used on the cover of an A5 pamphlet of activities for under-twos

Below: Hatched lines used to give tone – a useful technique to use when solids are, as in stencil duplicating, difficult to print (A4, stencil duplicating)

Pencil drawing screened into halftone for the front cover of a magazine (A4, offset, two colours)

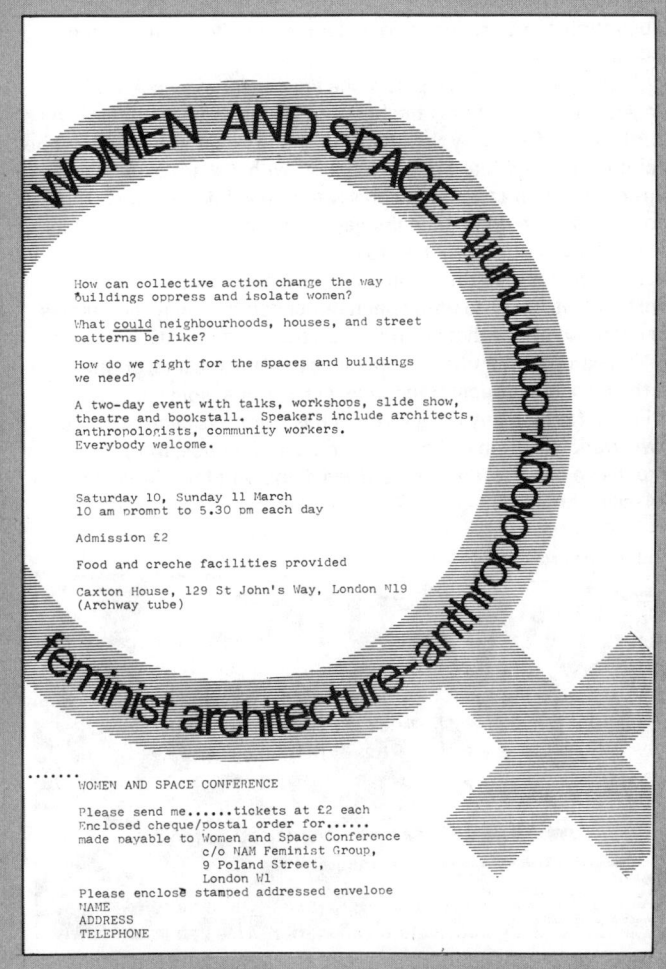

Instant tone used to create a background shape on this conference publicity leaflet (A4, offset)

OTHER TYPES OF IMAGE

COLLAGE
Photos, drawings or other images can be cut up and pasted together to make collages. These can be very effective if done well: they enable you to create fantastic images to great effect. (You will need to screen photos separately before you make the collage, or else you can screen the whole collage once it is made.)

IMITATIONS
An amusing idea in making your design is to parody the style of another publication: the range of dry-transfer lettering enables you to imitate most lettering styles. (The Monty Python books, in which every page is a spoof, are an example of this.)

PHOTO-PRINTS

Photo-prints are made on a process camera. This is a large camera specially designed to copy photos and paste-ups on to paper or film. Prints are made on the camera on many different brands of material and paper, and this accounts for the variety of different terms for photo-prints such as 'bromides', 'photostats', 'PMTs' or 'Veloxes'. We have used the term 'photo-print' to avoid confusion.

Photo-prints are made when screening photos into halftone dots or converting them to line (⇨ page 97). But they can be very useful for other things too:
- Any part of your design can be enlarged or reduced on the camera as the print is made. For example, a headline can be enlarged so that it fits right across the column or page, or an illustration can be reduced so that it fits exactly into the space available.
- If you need to remove part of the image (for example, the background on a photo) the best and easiest way to do this is on a photo-print. Cut away the area with a knife, paint it out with white paint ('process white') or scratch it off with a sharp blade. Doing this needs great care, but it can save money if done at this stage: if not, it has to be done once the negatives have been made, which is expensive and not so easy to control.
- You can improve the quality of your artwork by making a photo-print from it. You can easily paint out or scratch away dirty marks or paste-up edges that reproduce on the photo-print.

You can get your photo-print made at plate-makers, photo-setters, design groups, community presses or at some commercial printers. (In fact, any company that has a process camera for its own work will often make photo-prints as a service, too.) It is an expensive process, costing between £2·50 and £5 ($5–$10) for an A4-size photo-print.

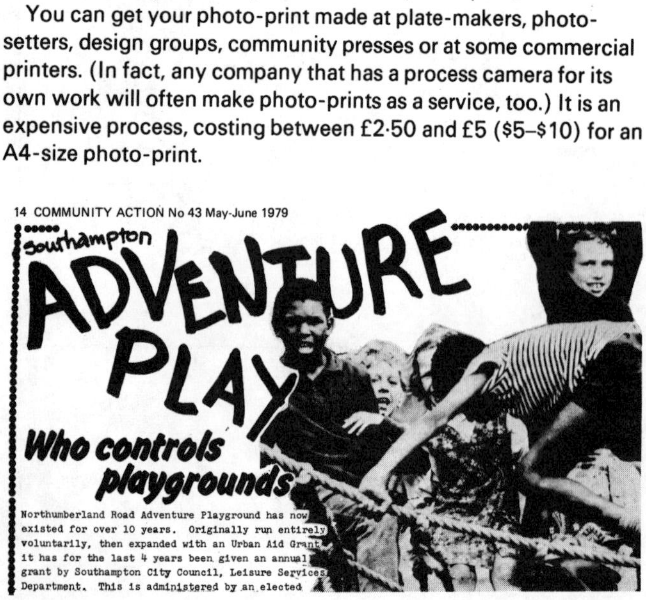

A screened photo-print with the background cut away, on a magazine page (A4, offset)

Collage creates this powerful image on the front cover of a magazine (A3, offset, two colours)

A poster advertising a book about newspapers and education uses imitation convincingly (A3, offset)

MAKING THE PASTE-UP

When all the different components of your design are ready, it is time to stick them down in position, in other words, to make the paste-up.

EQUIPMENT AND MATERIALS
- *Drawing-board* (or any smooth flat surface, such as a Formica-top table). You can paste up simple jobs almost anywhere if you have to, but it is easier to do a clean and accurate job on a drawing-board.
- *Ruler and set square (triangle)/T-square/parallel rule.* To help you to stick things down square and straight.
- *Masking tape.* The best thing for taping paper to the drawing-board, since, if stuck down lightly, it will not tear the paper when removed. It can also be used to remove mistakes in dry-transfer lettering.
- *Cow gum or rubber cement.* Rubber glue used to stick paper down flat. Cow gum is applied with a plastic spreader and rubber cement is applied with a brush.
- *Cow gum rubber or rubber-cement pick-up.* You can buy a special rubber eraser to remove excess rubber cement or Cow gum. Or you can use a ball of dried-up rubber glue (you can make this yourself).
- *Art knife.* You will need a small sharp knife to cut out accurately. A scalpel sounds very medical but it is also the best tool for most paste-up cutting. There are other, specially made art knives, like the X-Acto No. 1 knife and the snap-off knife, the blade of which can be snapped off and replaced when blunt.
- *Scissors.*
- *Light-blue (non-reproducing) pencil or marker.* Used for drawing the grid and other guide-lines so that you can stick things down accurately in the right position. The lines can be left on the paste-up because the camera will not pick them up when the paste-up is photographed to make a plate or stencil.
- *Scrap paper (lots of it!).* On which to lay pieces of artwork when applying glue.
- *Cutting-board (or more scrap paper, a piece of glass or heavy card).* To rest artwork on when trimming it with the art knife.
- *Paper.* Any white paper will do for the paste-up, but a thin card or mounting or illustration board is best, as it keeps the paste-up in better condition, is easier to handle and can be sent through the post with less chance of damage. With simple jobs you can use a sheet of paper of the same size as the finished print, but usually it is best to use card or board at least 10 cm (4 in.) larger overall than the printing area. This leaves space for any instructions you need to give to the printer.

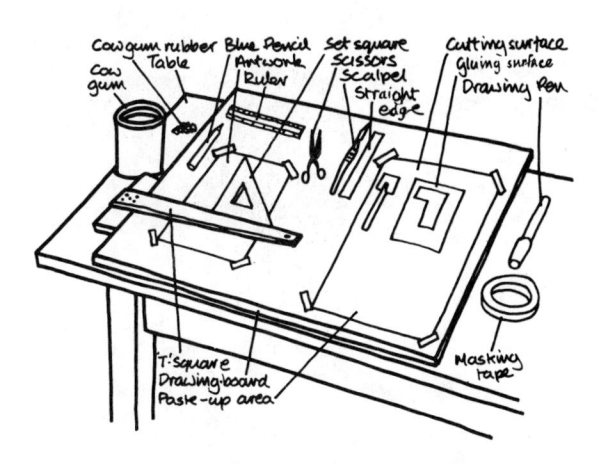

Equipment and materials for pasting up

DRAWING YOUR GRID
Unless you are using pre-printed grids (⇨ page 89), supplied by the printer or typesetter, you will need to draw out each page with the grid you have decided on. This provides clear guide-lines for you to stick things down straight.

1. Tape down the sheet of card or paper on the drawing-board with masking tape.

2. Using either a ruler and set square/triangle, or a T-square and set square/triangle, draw in the outside edges of the page in light-blue pencil. Make sure you are accurate and that the corners are square (with ninety-degree angles).

3. Draw in the columns and any other guide-lines you need (the top and bottom of the columns, the position of headlines, pictures and so on).

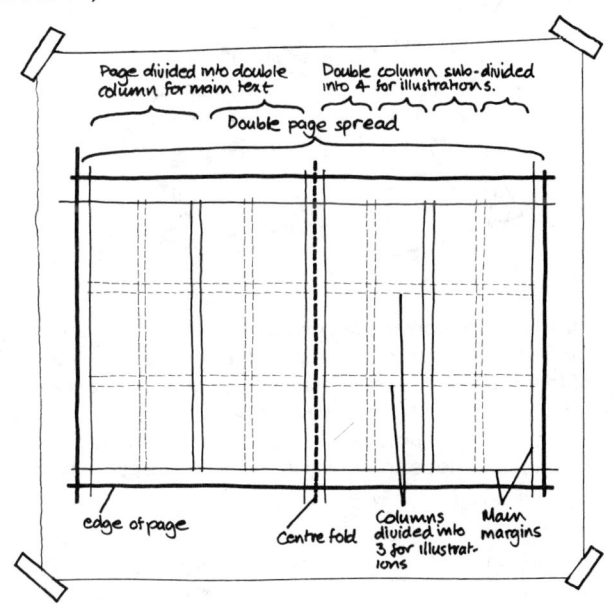

Marking up the page

4. To show the printer where the outside edge of the page is, make 'crop marks' in black pen at the corners as shown. These are to enable the printer to position the image correctly when making the plate and later when printing.

5. Mark any fold lines with a dotted line beyond the edge of the printed page, as shown.

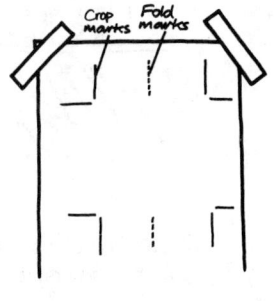

CHECKING THE LAYOUT
1. Trim away any excess paper from around the various pieces of your design (text, lettering, drawings, photo-prints etc.). Do not trim too close to the images unless it is absolutely necessary, because they will be damaged by glue if it gets on the top side; the edges may also show up as shadows on the plate or stencil and it will be very difficult to paint them out if they are very close to the image.

2. Check that everything fits in properly, by laying the pieces down on the paper in their correct positions. It is still possible to try out alternative arrangments of the design at this stage, even if it merely reassures you that you have chosen the best solution.

3. You may find it helpful to mark the positions of all the pieces (with your light-blue pencil), because you have to remove them from the paper before you start sticking them down.

PASTING UP
It is best to start pasting up at the top of the page and to work down – this avoids smudging or damaging the parts that have already been stuck down. If this is not convenient – for example, you may have to start by first gluing down the major component of the design in the middle, then fitting the smaller pieces in around it – cover the parts you have stuck with a sheet of tracing paper to protect them while you paste down the rest of the page.

How to stick things down

1. Lay the piece of artwork face down on clean scrap paper and apply a very thin coat of Cow gum or rubber cement with a plastic spreader or brush.

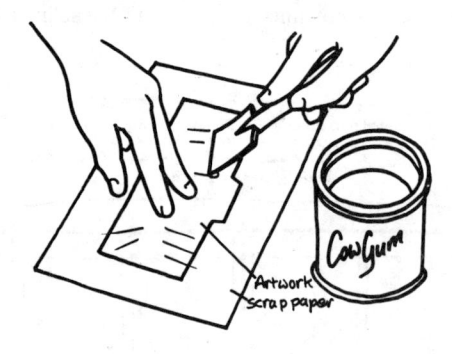

2. Lay the glued artwork in position on the paste-up and check with the T-square or set square/triangle that it is lined up with the edge of the column and parallel across the page. The glue takes about twenty minutes to dry, so you can slide the artwork around until it is in the correct position. Do this carefully with the pointed end of the scalpel, not a finger, which may mark the paper or lift bits of lettering.

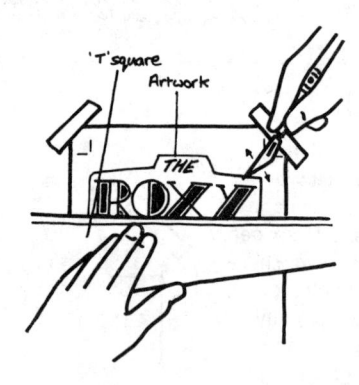

3. When the piece is in position press it down more firmly with the side of your hand, having first laid a sheet of clean paper on top to protect it.

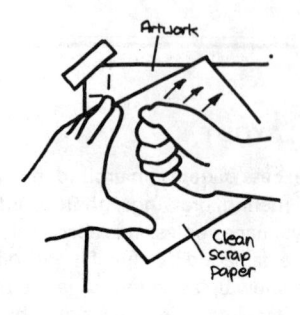

4. Clean off any gum that has seeped out from around the edges of the pieces with a Cow gum rubber or rubber-cement pick-up, once everything is stuck down and is dry. Do this carefully or you may smudge or remove lettering or typing if some glue has got on to them. If you do not clean up the artwork in this way the bits of glue or dirt that stick to it may show up on the print.

5. Protect the paste-up by taping a sheet of tracing paper or layout paper across it and fixing it at the back. This keeps the paste-up clean; you can also write instructions on it for the printer.

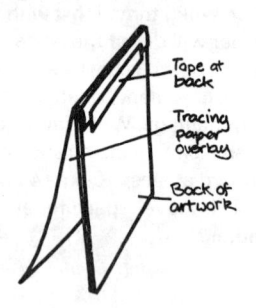

Other things you may need to put on the paste-up include the position of photos, instructions for colour printing and overlays for tone and reversed-out material.

POSITION OF PHOTOS
Indicate in light-blue pencil on the paste-up the positions of any photos that are being screened by the printer (⇨ page 98). If there is more than one photo, number each one on the back, number the space allotted to it on the paste-up, and on the back of the photo write the size to which it is to be reduced or enlarged.

PASTE-UPS FOR MORE THAN ONE COLOUR
A separate plate or stencil has to be made for each colour that you use. However, this does not always mean that you have to make separate paste-ups for each of them. If the colours are quite well apart on the page (as, for example, with a heading in colour) it is usually possible to make only the one paste-up, indicating on a tracing-paper overlay which parts are to be in which colours. The printer can then separate these at the negative stage (check beforehand that this is OK with your printer).

On jobs in which the colours interlink or overprint, you will need a separate piece of artwork for each colour. This is done on transparent overlays.

1. Make the paste-up for the main colour in the usual way on paper, and draw small crosses at the top and bottom of the page (outside the print area). These are called 'registration marks' and act as a guide so that the printer can line up the crosses on the different pieces of paste-up to make sure that they fit together accurately during printing.

2. Position a piece of tracing film (or acetate or tracing paper) on top of this first part of the paste-up and make registration marks in the same positions as the ones underneath. Paste down on the tracing film the parts of the design that are to be printed in the second colour, lining them up with the first colour, which you can see through the tracing film. (If two colours touch, allow a 2 mm/$\frac{1}{16}$ in. overlap.)

3. Repeat this process for each colour in the design and indicate on each overlay the colour that is to be printed in. (Either write the name of the colour, if you have chosen one from the printer's range, or attach a sample if you want the printer to mix up a special colour – which usually costs more!)

TONE OVERLAYS
If you want the printer to put on an area of tone at the negative stage (⇨ page 97), this should be indicated on a tracing-paper overlay showing the exact shape and position required. If it is a complicated shape cut it out of black paper and paste it in position on the overlay. Put registration marks on the paste-up and overlay (see above) to ensure the printer positions them correctly, and write instructions at the edge about the percentage of tone you require.

REVERSING OUT
An image can be photographically reversed to produce a negative, so that you get, for example, a white headline on a black background (such as the heading 'MAKING THE PASTE-UP' on page 103). This is called 'reversing out' (U.K.) or 'dropping out' (U.S.) and can be done to headlines, drawings, photos or areas of text.

Getting it done by the printer. Put the part of the design that you want reversed out on to an overlay (as if it were a second colour; ⇨ above) and write on the overlay that it is to be reversed out.

Doing it yourself. You can do the reversing yourself if you have access to a photographic dark-room (and this will cost you less than if you pay the printer to do it for you). For example, you can put dry-transfer lettering on to a piece of acetate and then contact print this on to a sheet of photographic paper to produce a white image on a black background. This is then laid down on the paste-up in the normal way.

Right: On this leaflet the grey tone of lettering in the background has been put on by the printer from artwork done on an overlay (A4, offset)

Teachers' Notes
The purpose of the game and accompanying background material is:

— to show students of all backgrounds that there are common problems which people of all backgrounds share, and that the question of 'race' is often used to divide people and to direct attention from these problems.

— to show students of all backgrounds how myths and prejudice lead to unfair discrimination against black and immigrant people,

— to encourage students to find out how people who are discriminated against can obtain their rights,

— to encourage students to value the contribution which people of all races and backgrounds can make to our society.

Glossary

Printers, designers and others who are professionally involved with print may confuse you with their jargon. In this book we have tried to avoid jargon, but to help make it clearer we have included in this glossary explanations of some of the more common words.

A sizes ⇨ *Paper sizes.*

Art paper. A high-quality glossy paper or card.

Artwork ⇨ *Paste-up.*

Ascenders ⇨ *Type.*

Binding. The fastening together of all the printed sheets to make the book, magazine etc.

Blanket. The rubber covering around the cylinder of an offset press which transfers the image from the plate on to the paper.

Bleed. The area of print that extends beyond the edge to be trimmed.

Bold. The heavier (blacker) version of a typeface.

Bromides ⇨ *Photo-print.*

Camera-ready artwork. A paste-up that is completely ready to be photographed.

Casting off. Estimating how many lines of type a piece of text will make when typeset in a given type size.

Character. Any individual letter, figure or punctuation mark in a typeface. (For casting-off purposes the spaces between words are also counted as characters.)

Chemical transfer. A plate-making method using a photocopying method to transfer the image on to the offset plate.

Collating. Putting printed sheets in the right order before binding.

Colour separation. The separation of a full-colour image into the four process colours with a process camera and filters or an electronic scanner.

Column inch. A measure used on newspapers based on a space one inch deep and a column wide.

Composition ⇨ *Typesetting.*

Continuous tone. A photo or image with a full range of tones.

Copy. (1) To reproduce with a photocopier. (2) The text to be printed.

Copy-fitting. Calculating the space the text will take up when typeset.

Cover paper. A range of heavy papers used for the covers of pamphlets, booklets etc.

Crop marks. Lines drawn on the paste-up to indicate to the printer where the printed sheet is to be trimmed.

Crown ⇨ *Paper sizes.*

Descenders ⇨ *Type.*

Dot for dot. The reproduction of an existing halftone by treating it as ⇨ *line.*

Drop out ⇨ *Reverse out.*

Dry-transfer lettering. Type that can be transferred to the paper by burnishing, such as Letraset.

Dummy. A mock-up of a page or of the whole publication to show how it will look.

Electronic stencil cutter. Also called a 'scanner'. Used to make stencils from artwork for ⇨ *stencil duplicating.*

Electrostatic method. A method of photocopying in which the image is transferred by being electrically charged by light. The latent image attracts ink which is then fixed to the paper.

Élite. A typewriter type size, with twelve letters to the inch.

Em. A typesetting measurement derived from the square of any given type size. The 12-pt em (the ⇨ *pica* em) is used for linear measurement of type.

En. Half the width of an em. As it is the average width of a character or space it is often used in costing typesetting.

Finishing. The processes that turn the printed sheets into a publication, e.g. ⇨ *collating,* ⇨ *trimming,* ⇨ *binding.*

Flat-bed press. A printing machine which prints from a printing surface that is lying in a flat plane.

Format. A general term for the size, shape and style of a publication.

Fount solution. The water solution in an offset press.

Full colour. The reproduction of a full range of colours by the use of four separate printing plates, including one for each of the primary colours. The four are magenta (red), cyan (blue), yellow and black, known as process colours.

Galley. In ⇨ *letterpress,* this is a tray of set type; the term is now widely used to refer to proofs of typesetting in long columns that have not yet been divided into pages.

G/m². (or **gsm**). Grams per square metre. A measure of the weight of the paper and an indication therefore of the thickness.

Gravure (or **photogravure**). A printing method in which the image is engraved into a copper ⇨ *plate.* Used for long-run full-colour printing, like that of colour supplements.

Grid. The framework of lines marking the margins and columns of a page.

Gripper-edge. The part of the paper at the edge which is gripped by the machine to feed the paper through the press.

Gutter. The space between the columns where two pages meet at the fold.

Haberule ⇨ *Type-scale.*

Halftone. The reproduction of continuous tone to obtain the tonal gradations by a pattern of minute dots of varying sizes. This is produced by 'screening'; that is, by photographing the image through a dot screen on a process camera.

Hot metal. Type which is set by being cast from hot metal.

Imperial ⇨ *Paper sizes.*

Imposition. Arranging pages that are to be printed on one sheet in such a way that they are in the right numerical order when

GLOSSARY

the printed sheet has been folded (and trimmed).

Impression cylinder. The cylinder in an ⇨ *offset* press which presses the paper against the ink image on the blanket cylinder and makes the print.

In-plant (printing). Printing by a unit within an office or organization.

Instant printer. A high-street print-shop specializing in quick photocopying and small offset work.

Jobbing printer. A small (often letterpress) printer who mainly handles small jobs like business cards, invitations and letterheads.

Justification. Typesetting lines of text so that each line is of the same length, so creating the straight-sided 'column' of type of regular width throughout.

Keyline. A line or a solid area of red or black which indicates on the artwork the position of photos to be screened by the printer. Also known as the 'holding line'.

Knocking up. Making a pile of paper square so that each sheet feeds easily through the press or duplicator.

Landscape. An oblong sheet of paper laid on its long side (the opposite to ⇨ *portrait*).

Layout. Plan for the arrangement of the print material (text, illustrations etc.) to show what it should look like when set and put together.

Leading (pronounced 'ledding'). The space between lines of type.

Letterpress. The traditional method of printing, using movable relief letters and blocks for illustrations.

Line. Any image for printing that consists of pure black and white and does not include greys or continuous tone.

Lithography. Printing method that exploits the incompatibility of water and grease. The most common form today is ⇨ *offset*-litho.

Logo. A printed device or letter design used as a symbol or trade mark by an organization.

Lower case. The small characters of the alphabet as opposed to the capitals.

Machine-glazed (MG). Paper coated on one side, used for printing posters.

Makeready. The work needed to prepare the press or plates before the job is actually printed.

Marking up. Writing the specifications on the text for a job that is to be typeset.

Measure. The width of a column of typesetting, measured in picas.

Mechanical ⇨ *Paste-up*.

Negative. A photographic image on film or paper in which the black becomes white and the white becomes black.

Newsprint. A cheap, thin, off-white paper used for printing newspapers.

Offset. Short for 'offset-litho'. So called because the ink image is transferred (offset) from one cylinder (the plate cylinder) to another (the rubber-blanket cylinder) on to the paper.

Opaque. Paint which does not allow light to pass through, used to paint out unwanted spots on film negatives or positives.

Organdie (or **organdy**). Fabric that can be used as a mesh for screen printing.

Overlay. Transparent paper flap fastened over paste-up to protect it and used to write instructions to the printer (about tones or colour separations etc.).

Overprint. Printing of one colour on top of another.

Paper sizes. The standard international metric paper sizes are superseding the traditional British paper sizes (although you may still come across the latter sometimes). The international sizes are based on three series: A, B and C. The most common of these is the A series:
A1   594 × 841 mm ($23\frac{3}{8}$ × $33\frac{1}{8}$ in.)
A2   420 × 594 mm ($16\frac{1}{2}$ × $23\frac{3}{8}$ in.)
A3   297 × 420 mm ($11\frac{3}{4}$ × $16\frac{1}{2}$ in.)
A4   210 × 297 mm ($8\frac{1}{4}$ × $11\frac{3}{4}$ in.)
The RA series is slightly larger, to allow for trimming down to A sizes:
RA1   610 × 860 mm ($24\frac{3}{8}$ × $34\frac{3}{8}$ in.)
RA2   430 × 610 mm ($17\frac{1}{8}$ × $24\frac{3}{8}$ in.)
The SRA series is slightly larger still, to allow for trimming 'bleed' edges:
SRA1   640 × 900 mm ($25\frac{5}{8}$ × 36 in.)
SRA2   450 × 640 mm (18 × $25\frac{5}{8}$ in.)
Traditional British sizes. Of these, the most likely sizes you will come across are:
Crown   15 × 20 in.
Double crown   20 × 30 in.
Imperial   21 × 23 in.
U.S. sizes

Base sizes	Usual sizes	
17 × 22 in.	26 × 40 in.	$8\frac{1}{2}$ × 14 in.
25 × 38 in.	25 × 38 in.	$8\frac{1}{2}$ × 11 in.
20 × 26 in.	23 × 35 in.	$9\frac{1}{2}$ × $12\frac{1}{2}$ in.
	23 × 29 in.	10 × 13 in.
	$17\frac{1}{2}$ × 22$\frac{1}{2}$ in.	11 × $14\frac{1}{2}$ in.

Paper weights. Paper is graded by weight, which provides an indication of the thickness of the paper.
Metric paper weights are measured in grams per square metre (g/m² or gsm). Weights commonly in use are 60, 75 and 80 g/m².
In the U.S.A. the weight of 500 sheets of the base size of a particular type of paper is the unit of measurement. For offset, book and text paper the sheet size is 25 × 38 in., for bond, duplicator and ledger it is 17 × 22 in. and for cover it is 20 × 26 in. Weights normally used are 50, 60 and 70 lb.

Paste-up. Drawings, illustrations, text etc., pasted down in the correct position ready for reproduction. Also called the 'artwork', 'original' or 'mechanicals'.

Perfect binding. A binding method in which the pages are held together and fixed to the binding with glue.

Photo-print. A photographic print, made on a ⇨ *process camera*, of the paste-up or part of the paste-up. Also called 'PMT', 'Velox', 'stat', 'Photostat' and 'bromide'.

Pica. (1) A unit for linear measuring in typesetting (really a pica ⇨ *em*), e.g. for the length of line or margins. There are twelve points in a pica and six picas in an inch.
(2) A typewriter type size, with ten letters to an inch.

Pinholes. Small imperfections in a film negative or positive which allow light to pass through. Must be painted out (⇨ *opaque*).

Plate. The surface which has been treated to carry an image – made from various materials, including paper, plastic and metal.

Platen. A letterpress printing machine that presses two flat surfaces together to make the print.

Point. The standard unit for measuring type. There are seventy-two points to an inch.

Portrait. An oblong sheet of paper laid on its short side (the opposite to ⇨ *landscape*).

Positive. A photographic reproduction on paper or film in which the black remains black and the white remains white (as opposed to a negative, where they are reversed).

Print and turn. A procedure for printing both sides of the paper from one plate or stencil.

Process camera. Also called a 'copy camera'. Large camera designed for process work: making negatives the size of the finished print for plate-making, screening photos, copying artwork, enlarging and reducing on to film or paper (⇨ *Photo-prints*).

Proof. A trial print that is checked against the original text for errors. Page proofs are also used to check that a paste-up has been correctly positioned.

Proofing press. A flat-bed letterpress machine used to take a ⇨ *proof* print. Can also be useful for short runs of posters.

Proportional spacing. In typesetting each letter has a varied width, unlike typewriting, in which each letter is of the same width.

Ragged ⇨ *Ranged*.

Ranged (or **Flush**) **left/right.** Typesetting in which the lines are aligned (vertically) on the left or on the right.

Ream. Five hundred sheets of paper.

Registration. Making sure that the image is printed in the correct position so that successive printings coincide exactly.

Registration marks are small crosses, on the paste-up for each colour, which are lined up to ensure perfect registration.

Relief printing. A method that prints from a raised image area.

Reprographics. A general term for newer printing methods such as microfilming, spirit duplicating, ⇨ *stencil duplicating*, photocopying and small offset printing.

Reverse out. To reverse the black-and-white areas of any image. Can be done by the printer at the negative-making stage, or by using a special photographic paper on a ⇨ *process camera*.

Rules. Lines, such as those of borders or 'boxes' around the text, are called rules. (They are not called lines: a line means a line of type!)

Run. The number of copies to be printed.

Run-on price. The cost of an extra number of prints on top of the initial print run in a quote.

Scanner ⇨ *Electronic stencil cutter.*

Screen ⇨ *Halftone.*

Screen printing. A printing method which uses a frame (the screen) over which is stretched a fine mesh on which a stencil is stuck. Also called 'silk screen', 'serigraphy' and 'screen process printing'.

Set-off. Unwanted transfer of wet ink from one printed sheet to the next.

Show-through. Printing showing through from one side of the paper to the other.

S/S. Abbreviation for 'same size' (used of material for reproduction that is to be neither enlarged nor reduced).

Stencil duplicating. Simple printing process for short runs, using a stencil master. Also called ink duplicating, cyclostyling, Roneoing, Mimeoing, Gestetnering etc.

Stitching. Trade term for stapling.

Stripping in. Inserting screen negatives in their correct position on a line negative.

Thermocopying. A copying method which uses heat as opposed to light (as in photocopying). Can be used to make stencils for ⇨ *stencil duplicating* and masters for spirit duplicating.

Thermography. A process whereby special inks are heated after printing to produce a relief effect.

Tones. A pattern of dots that gives the effect of greys. Can be produced by using 'instant tone', which comes in a variety of patterns and densities, or by 'screening'.

Trimming. Cutting the paper to size after printing.

Type. Characters used in place of the (handwritten or typed) manuscript to produce the printed text. The main features of the type 'face' are:

Type-scale. A special ruler marked in point sizes, used to measure the depth of a given number of lines of typesetting.

Typesetting. The typing out or assembling of type for printing. Type can be set by hand or by machine.

Unit price. The price of a single copy of what you are printing or selling.

Unjustified. Text setting in which the column of lines is straight on one side and irregular on the other.

Upper case. Capital letters.

Varityper. An older sort of proportional-spacing typewriter.

Web printing. Printing from a press which is fed paper from a reel rather than from sheets.

Useful Books

Screen printing

R. Cuff and P. Cartwright, *Screen Printing*, London, Nelson, 1975

Stephen Russ, *Practical Screen Printing*, London, Studio Vista, 1972

Stephen Russ, *Setting Up in Screen Printing*, Bath, Bath Academy of Art/The Society of Education Through Art, 1977

Chris Treweek, *How to Screen Print*, London, Macmillan, 1983. (Wall charts)

Silvie Turner, *Screen Printing Techniques*, London, Batsford, 1976; New York, Taplinger, 1979

Offset-litho printing

Clifford Burke, *Printing It*, Berkeley, Calif., Wingbow Press, 1976

Joseph Stellar, *Trouble Shooter on the 1250 Multi*. Available from: Reliable Duplicator Service, 75707 Sanford Sta., Los Angeles, Calif., CA 90005/Blackrose Press, 30 Clerkenwell Close, London EC1

U.S. Navy, *Lithographer 3 & 2*, P.O. Box 1533, Washington DC 20013, U.S. Government Printing Office (the U.S. Navy's manual on offset)

Relief printing and letterpress

J. Erickson and A. Sproul, *Printmaking without a Press*, London, Van Nostrand Reinhold, 1974

Peter Green, *Introducing Surface Printing*, London, Batsford, 1967

J. Ross and C. Romano, *Complete Relief Print*, Collier-Macmillan, 1974

Herbert Simon, *Introduction to Printing*, London, Faber, 1968

Type and Typesetting for Users of Adanas, London, Adana

Harry Whetton (ed.), *Practical Printing and Binding*, London, Odhams, 1946

Photocopying

Patrick Firpo, Lester Alexander, Claudia Katayanagi and Steve Ditlea, *Copyart*, New York, Richard Marek, 1978

Design and paste-up

David Cherry, *Preparing Artwork for Reproduction*, London, Batsford, 1976

Ken Garland, *Graphics Handbook*, London, Studio Vista, 1972

Ken Garland, *Illustrated Graphics Glossary*, London, Barrie & Jenkins, 1980

Jan Goodchild and Bill Henkin, *By Design: A Graphics Source Book of Materials Equipment and Service*, New York, London, Tokyo, Quick Fox, 1980

Bill Gray, *Studio Tips*, London, Van Nostrand Reinhold, 1976

John Lewis, *Typography: Design and Practice*, London, Barrie & Jenkins, 1980; New York, Taplinger, 1978

Cal Swann, *Techniques of Typography*, London, Lund Humphries, 1969

General printing

Richard J. Broekhuizen, *Graphic Communications*, Bloomington, Ill., MacKnight, 1979

James Craig, *Production for the Graphic Designer*, London, Pitman, 1974

Armin Hofmann, *Graphic Design Manual*, London, Van Nostrand Reinhold, 1965

Kodak, *Graphic Arts Handbook*

Leslie H. May, *Printing Reproduction Pocket Pal*, London, Advertising Agency Production Association, 1969

David J. Plumb, *Design and Print Production Workbook*, London, Workbook Publications, 1978

Silvie Turner, *Handbook of Printmaking Supplies*, Printmakers Council of Great Britain, 1977

Jonathan Zeitlyn, *Print: How You Can Do It Yourself*, London, Inter-Action Inprint, 1980

Publishing media

Hans Magnus Enzensberger, 'Constituents of a theory of the media', *The Sociology of Mass Communication*, London, Penguin Books, 1972

Harold Frayman, David Griffiths and Chris Chippindale, *Into Print: A Guide to Non-Commercial Newspapers and Magazines*, London, Teach Yourself Books, 1975

Denis MacShane, *Using the Media*, London, Pluto Press, 1979

For inspiration

Berjouhi Barler, *The World as Image*, London, Studio Vista, 1970

Szymon Bojoko, *New Graphic Design in Revolutionary Russia*, London, Thames & Hudson, 1972

For kids

Harvey Daniels and Silvie Turner, *Simple Printmaking with Children*, London, Van Nostrand Reinhold, 1972

Howard Mell and Eric Fisher, *I Can Do It – Printing*, Huddersfield, Schofield & Sims, 1968

Organizations

U.K.

Association of Little Presses
262 Randolph Avenue
London W9
(Poetry publishers)

Association of Print Workshops
Printmakers Workshop Ltd
29 Market Street
Edinburgh EH1 1DF
(Mostly fine-art print workshops)

British Printing Society
Hon. Secretary
William F. Coles
725 Old Kent Road
London SE15 1JL
(Mostly amateur printers)

Publications Distribution Cooperative
27 Clerkenwell Close
London EC1R 0AT

U.S.A.

American Association of Independent Publishers
238 N. Juanita Ave.
Los Angeles
California CA 90004

New York State Small Press Association
Box 1264
Radio City Station
New York
NY 10101

Acknowledgements

We should like to thank all the groups (listed below) that have given us permission to use their work as examples in this book. Every effort has been made to trace copyright-holders and the publishers would be pleased to hear from any not acknowledged here.

Anarchist Communist Association
Brent Publications
British Printing Society
Camden Anti-Nazi League
Camden Square Playground
Camerawork
Campaign against Racism and Fascism
Canonbury School Parent–Teacher Association
Caxton House Mother and Toddler Group
Community Action
Community Press
Community Service Volunteers (*Hassle* teachers' notes)
Corner House Bookshop
Cromer Street Women's Centre
Crumbles Castle Playground
Drayton Park Toy Library
The Factory
Fleet Community Education Centre
Freightliners Farm
Hackney and Islington Branch/National Childbirth Trust
Hackney Gutter Press
Hackney People's Press
Hackney Under-Fives
Haringey Branch/National Childbirth Trust
Highbury Roundhouse
Holloway Neighbourhood Group (*design*: Sean McGary)
Inter-Action Inprint (*Community Newspapers*)
Islington Archaeology and History Society
Islington Christians against Racism and Fascism
Islington Gutter Press
Islington Nalgo
Islington Play Association
Islington Preschool Playgroups Association
Islington Training Unit
Islington Under-Fives Group
Ital Rockers
Kids' Stuff magazine
KPE Printers
Kropotkin's Lighthouse
Labour Newspaper Group
Lavender Hill Family Workshop
The Leveller
London Adventure Playgrounds Association (trainees)
Mansfield Neighbourhood Festival
E. A. Markham (Games & Penalties)
Mary Ward Centre
Mayville Community Centre
Ian Mortimer
National Playbus Association
New Architecture Movement (Feminist Group)
North London College
North Road Playground
Ormond Road Workshops
Paddington Printshop
Peace News
People's Field Day
Pooles Park Tenant Co-op
Poster Collective
Prisoners' Aid Committee
Pronk, Davis & Rusby
Redesign
Rising Free Bookshop
Russell Press
St Pancras United Tenants Association
See Red
Socialist Workers Party
Society of Analytical Psychology
Swapo
Temporary Hoarding
Thornhill Project/Copenhagen School
Undercurrents (*artist*: David Hall)
Upper Street Under-Fives Centre
Women in Manual Trades
Women's Arts Alliance
Wooden Bridge Playground